in' no how the
running

Don't Never Tell Nobody Nothin' No How

RICK JAMES

Don't Never Tell •••• NOBODY NOTHIN' NO HOW

THE REAL STORY OF WEST COAST RUM RUNNING

HARBOUR
PUBLISHING

HARBOUR PUBLISHING CO. LTD.
P.O. Box 219, Madeira Park, BC, VON 2HO
www.harbourpublishing.com

Edited by Peter Robson
Indexed by Emma Skagen
Jacket design by Anna Comfort O'Keeffe
Text design by Shed Simas / Onça Design
Maps by Roger Handling
Printed and bound in Canada
Printed on FSC-certified paper

Harbour Publishing Co. Ltd. acknowledges the support of the Canada Council for the
Arts, which last year invested $153 million to bring the arts to Canadians throughout
the country. We also gratefully acknowledge financial support from the Government of
Canada and from the Province of British Columbia through the BC Arts Council and the
Book Publishing Tax Credit.

LIBRARY AND ARCHIVES CANADA CATALOGUING IN PUBLICATION
James, Rick, 1947-, author
    Don't never tell nobody nothin' no how : the real story of West Coast rum running /
Rick James.

Includes bibliographical references and index.
Issued in print and electronic formats.
ISBN 978-1-55017-841-8 (hardcover).--ISBN 978-1-55017-842-5 (HTML)

    1. Alcohol trafficking--British Columbia--Pacific Coast--History--20th century. 2.
Prohibition--British Columbia--Pacific Coast--History--20th century. I. Title. II. Title: Do
not never tell nobody nothing no how.

HV5091.C3J36 2018   364.1'3361097111   C2018-903485-8
                                                  C2018-903486-6

The Two Jims, wherever they are

TABLE OF CONTENTS

DON'T NEVER TELL NOBODY NOTHIN' NO HOW

L ooking back, it seems that I was destined to write about West Coast rum running. My dad, Dick James, was born in the Shetland Islands to Richard Pascoe James, a Cornish tin miner, and Ruby (née Scott). Granny's family had made a living from the sea for generations and her grandfather and father, both named Peter, supplemented their hardscrabble existence as cod fishermen and whalers by smuggling tobacco and spirits. As it was, smuggling was never considered a crime or a sin by Shetlanders who struggled to get by on the isolated Scottish archipelago off the far end of the North Sea.

William Smith, a merchant and fish curer in the port village of Sandwick, who crewed with Ruby's grandfather on the packet boat *Rising Sun* in 1891, recalled that Scott Jr. (Ruby's father), said his father didn't really care for strong drink and so seldom got into positions that he could not extricate himself from. At the time Scott Sr. was running *Rising Sun* as a "Cooper," or smuggling vessel. They would fill the hold with tobacco and spirits at ports in Holland and Germany and then run the cargo across to the Yarmouth, Hull and Grimsby fishing fleets where they carried on a brisk trade. Prior to *Rising Sun*, Scott Sr. ran *Martha* of Geestemünde with son Peter as mate. The Scotts acquired the German boat after she was caught smuggling off the Shetland Islands coast in 1886. After they pleaded guilty, master and owner were fined twenty pounds sterling or the alternative of serving thirty days' imprisonment, while the three-man crew was fined five pounds each or twenty days in prison. Boat and cargo were subsequently seized and the boat ended up in the hands of the Scott boys.

Four generations later, the appeal of earning a living smuggling came to the fore in my own life. After graduating from Oak Bay High School in Victoria in the mid-1960s, smoking pot and hash was a recreational activity the crowd I hung out with often indulged in. We were in our late teens and early twenties, still trying to sort out how to navigate the adult world and preoccupied with how to get by or, at least, earn a half-decent income without resigning ourselves to a boring, humdrum job around town. Some who were adventurous enough headed up island to work in the woods setting chokers or landed jobs on trollers as deckhands. But living on the south end of Vancouver Island, where it's only a few short miles across Haro Strait into United States waters and the San Juan Islands, there was another, riskier enterprise that offered the potential of a very lucrative reward. And a few of my pals were crazy enough to try it.

It only required finding a reasonably reliable boat with a good, fast engine and making a run across Haro Strait, preferably during a dark and overcast night, to pick up an order of a few pounds of "product" on the other side of the line. Of course, it was all highly illegal but that's how economics works: the higher the risk, especially with the selling of a banned and illegal substance, the greater the potential for an exceptional return on one's investment. The idea seemed so very straightforward when presented by a couple of my colleagues who were always pressuring me to take part in one of their sketchy money-making schemes.

Forty years later, one old friend finally opened up and divulged his secret to success. To begin with, since my crowd was mostly from long-time Victoria families, they either kept their boats at the Oak Bay Marina or trailered them down to the Cattle Point boat launch. Then on a good dark night with no moon in the sky, they would race across into Washington state waters to Deadman Bay on San Juan

Island. This particular location on the chart was Vancouver Island drug runners' preferred spot for making a pickup. First of all, this rather quiet and isolated bay lies just outside of San Juan Island's Lime Kiln Point State Park, and Lime Kiln Lighthouse serves as an excellent navigational aid for those running across the strait from Victoria. Also, Deadman Bay is reasonably well sheltered with a nice moderately sloped beach for pulling a small boat up on. But not only that, the island's West Side road comes almost right down to the water in the bay. Here, their Canadian partners in the venture car or truck would arrive after picking up some pot in Oregon, so-called cheap Mexican shit selling for around seventy to seventy-five dollars a pound back in those days. After the transport vehicle arrived down at Deadman Bay, the product was discharged and loaded into the boat, which then raced off across the strait into Canadian waters. Once the weed hit the streets of Victoria, the dealers were able to sell it for a hundred and twenty dollars a pound.

Another reason why Deadman Bay worked so well was that there was little in the way of law enforcement on San Juan Island in those halcyon days of the 1960s and 1970s. As my old bud pointed out, there was virtually no US Coast Guard around in those waters, and he didn't recall there ever being any border protection service, drug enforcement or customs agents about, especially on the Haro Strait side of the island. The only police presence there at the time was the local sheriff's office. Even so, a number of the crowd I hung out with didn't have the sense to know when to call it quits. While many of us got into psychedelics, some took it a little too far and not only got into coke but even proceeded into far worse intoxicants like speed (methedrine) and junk (heroin) and bore the consequences. Then there were those who got busted for possession or, worse yet, running drugs into the country, and ended up with a record or even jail time. Even though many were actually quite bright and charming individuals, once they spent time behind bars they were never quite the same.

I always wondered how these jokers thought they could continue to pull off these hare-brained schemes in the first place. It all started to make sense once I became immersed in research for this book and learned about the escapades of numerous West Coast characters who turned to rum running while Prohibition south of the border made a futile attempt to dry out the American citizenry. The Noble Experiment, as it was so aptly branded, remained in effect from 1920 through 1933 and declared illegal the manufacture, sale, importation and transportation of alcohol throughout the US and, of course, the imbibing of such products. The parallels were intriguing and the stories very similar to what was happening throughout BC's south coast waters forty to fifty years later when local government authorities and law enforcement were attempting to put a stop to the burgeoning drug trade. Indeed, although we were unaware of it at the time, "jumpin' the line" (the border) was the commonly used phrase back in the 1920s to describe the actions of those gutsy enough to try to make surreptitious runs across Haro Strait to the San Juan Islands or even all the way down into Puget Sound. But back then, the particularly adventurous were filling their holds and packing the decks with case upon case of good Scotch, brandy and bourbon rather than high-test weed, hashish or LSD.

Probably the most surprising revelation that occurred while researching this book was that, contrary to what most of us have been led to believe, not all smuggling during the Prohibition years was marked by violence. While the criminal element was definitely involved with smuggling and bootlegging throughout the US, where hijacking and violent shootouts were the order of the day, when it was transpiring in British Columbia waters or out in international waters off the US coast, the activity was all carried out in a very civilized and politely Canadian manner. Rather than an adrenalin-filled and dangerous undertaking dominated by hoodlums, it was, on the whole, just another export shipping enterprise working out of the ports of Vancouver and Victoria and run by a number of generally charming and professional mariners and businessmen. Still, contrary to what many of these individuals may have convinced themselves and would have us believe, there was indeed the odd shootout and even one particularly gruesome murder that occurred in the Gulf Islands on the Canadian side of the line.

Overall though, the generally decent approach Canadians took to rum running during US Prohibition was very much like what we experienced in BC back in those carefree days when my old crowd was preoccupied with importing relatively harmless soft drugs and psychedelics across the line. For instance, I personally don't recall any of my buds packing firearms; that only came about when the consumption

The author selling the radical *Georgia Straight* rag at the corner of Yates and Douglas Streets in Victoria, circa 1969. Space Pals photo, Rick "Lou Lemming" James collection.

of the hard stuff, like meth, coke and heroin, became more widespread sometime around the mid-1970s. Another surprise was how many of the respected elderly citizens of Victoria and Oak Bay were active players involved in rum running. So I wonder if jumpin' the line has always been part of our local Vancouver Island culture—where those who became such dedicated participants in those wild and freewheeling years of the late 1960s and early seventies had somehow been socialized to pick up where some of the older residents in our community had left off.

Chapter One

SCOTCH OASES IN A DESERT OF SALT WATER

At the stroke of one minute past midnight, January 17, 1920, the proposed Eighteenth Amendment to the United States Constitution, the National Prohibition Act, known informally as the Volstead Act, was officially declared in effect. Named for Andrew Volstead, the Republican chair of the House Judiciary Committee who managed the legislation, the act stated that from that day on, "no person shall manufacture, sell, barter, transport, export, deliver, or furnish any intoxicating liquor except as authorized by this act." The Noble Experiment was to last fourteen years before being brought to an end in December 1933.

However, while the American government was closing tavern doors, the citizens of British Columbia chose to take a different direction that very same year. In British Columbia, a bill banning the sale of liquor, except for medicinal, scientific, sacramental and industrial purposes, was initially approved by referendum during the September 1916 provincial election, with Prohibition going into effect in September the following year. Right from the start, the attempt to prohibit the consumption of alcoholic beverages met with limited success. The legislation was not only extremely unpopular but in short order the government found it was difficult and expensive to enforce. It quickly proved a failed experiment. After only three years, a plebiscite was put to British Columbia voters (which for the first time included women) in October 1920, which read, "Which do you prefer? 1. The present Prohibition Act? 2. An act to provide for the government control and sale in sealed packages of spirituous and malt liquors?" It wasn't any surprise when the residents of the province voted thumbs down on the existing legislation. By June the following year it was all officially over and done with, and the very unpopular legislation slipped away into the proverbial dustbin of history. Each of Canada's nine provinces and two territories had experimented with Prohibition law but nearly all had repealed it by the late 1920s.

Still, the doors to BC's drinking establishments weren't exactly thrown wide open to thirsty residents looking for relief with an alcoholic beverage. Instead, what voters had approved in the referendum was a system of strict government control of the sale of alcohol with a three-man Liquor Control Board (LCB) to be set up to oversee and regulate the sale through government stores. Also, the new liquor act initially banned all public drinking unless one had a special permit issued by the LCB. This required the purchase of an annual liquor permit for five dollars and sales were limited to those twenty-one years of age and older. Additionally, BC residents weren't able to buy liquor by the glass until beer parlours were finally opened in 1925 following an amendment to the Government Liquor Act. These establishments, with their separate men's and ladies' entrances, soon became popular watering holes throughout the province, where they were to retain a prominent place in BC culture as a place to relax and unwind.

The Volstead Act was not only enacted the year British Columbia turned its back on Prohibition but also arrived in the midst of a serious postwar depression. Following the signing of the armistice in November 1918, a booming wartime economy turned stagnant and ground to a standstill. Much to the distress of soldiers returning from the Western Front, there were no jobs and they found themselves only contributing to a growing unemployment problem. As a result, many displaced workers and their families throughout British Columbia, as well as the rest of the nation, were looking for

A flood of liquor from Canada across the border into the United States was already underway by 1920. While Canada's federal government had outlawed the manufacture, transportation and sale of beverage alcohol, it could still be sold for medicinal, scientific, industrial and sacramental purposes. These provided the loopholes which bootleggers and smugglers were quick to take advantage of.
"By Jing, the Old Ceiling Leaks." Morris for the George Matthew Adams Service: *The Literary Digest*, 1920.

any means to get by. And some soon proved cleverer than others, especially those who happened to own anything that could float. While Prohibition had actually been in effect in the state of Washington since 1916, it was only when it was declared nationwide in 1920 that many on both sides of the border were quick to take advantage of this curious juxtaposition of Canadian and American government policies concerning alcohol. In the opinion of a large section of the public, the consumption of alcohol wasn't really considered a crime as such, much like the recreational smoking of marijuana is viewed today. As a result, in the early years of the 1920s, bootlegging (smuggling and delivering up liquor by land) and rum running (smuggling by water) quickly developed into an extremely lucrative enterprise throughout southern British Columbia.

Once the Volstead Act went into effect, it didn't take long before fleets of vessels, from weather-beaten old fishboats to large ocean-going steamers, began filling their holds with liquor. They would sail south from Canadian ports and sit offshore in international waters just outside the US territorial limit to deliver up their much-valued cargos to launches running out from shore. British Columbia was perfectly situated for the movement of illegal liquor by sea, particu-larly in local Canadian waters.

The southern tip of Vancouver Island was ideally positioned since it sticks out like a boot kicking into the exposed butt of Washington state. With a maze of islands scattered throughout the deep, sheltered waterways of the Strait of Georgia and Haro Strait, and the ports of Vancouver and Victoria only a short distance away from the American San Juan Islands, the area soon proved a verit-able floating liquor marketplace where Canadian boats delivered up orders to their American counterparts in relative safely.

Still, some rum runners, especially those who custom built their own high-powered speedboats, were willing to take it a step further and reap a better monetary return by making a run under the cover of darkness to deliver up their payload into Washington state beaches. Regardless of how it was carried out, the underground trade proved extremely profitable and by 1924, it was estimated that some five million gallons of booze had been smuggled by land and sea into the US. Bootleggers and rum runners were simply taking advantage of the basic law of supply and demand.

But the scale of rum running out of Canada's West Coast ports of Vancouver, Victoria and Prince Rupert remained relatively minor compared to that which was underway along the eastern seaboard

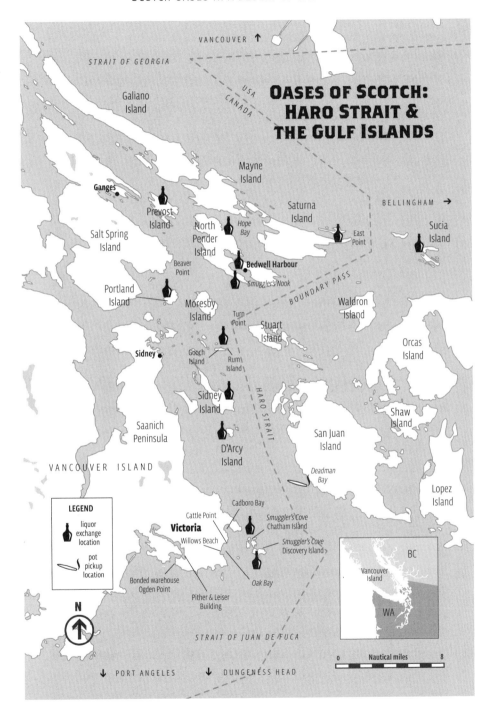

OASES OF SCOTCH: HARO STRAIT & THE GULF ISLANDS

VANCOUVER ↑

STRAIT OF GEORGIA

USA
CANADA

Galiano Island

Mayne Island

Saturna Island

BELLINGHAM →

Ganges

Prevost Island

Hope Bay

North Pender Island

East Point

Sucia Island

Salt Spring Island

Beaver Point

Bedwell Harbour

Smuggler's Nook

Portland Island

Moresby Island

Turn Point

Stuart Island

Waldron Island

BOUNDARY PASS

Orcas Island

Sidney

Gooch Island

Rum Island

Sidney Island

Saanich Peninsula

D'Arcy Island

San Juan Island

Shaw Island

HARO STRAIT

VANCOUVER ISLAND

Deadman Bay

Lopez Island

LEGEND

liquor exchange location

pot pickup location

Cadboro Bay

Cattle Point

Victoria

Willows Beach

Smuggler's Cove Chatham Island

Smuggler's Cove Discovery Island

Bonded warehouse Ogden Point

Oak Bay

Pither & Leiser Building

Vancouver Island

BC

WA

N

STRAIT OF JUAN DE FUCA

0   Nautical miles   8

↓ PORT ANGELES   ↓ DUNGENESS HEAD

and down in the Caribbean. While Halifax, the French islands of Saint Pierre and Miquelon lying off the south coast of Newfoundland, Havana, and Nassau in the Bahamas all served as major entrepôts, some four-fifths of illegal liquor was estimated to have entered the US through the Windsor–Detroit area on the Great Lakes, according to Dave McIntosh, author of *The Collectors: A History of Canadian Customs and Excise.* In his detailed history of the United States Coast Guard activities during the Prohibition years, *Rum War at Sea*, retired commander Malcolm F. Willoughby devoted only one chapter to the liquor trade along the Pacific coast. Overall though, estimates of Canada's share range from 60 to as much as 90 percent of the contraband market that was flooding into the US both by land and sea.

Of course, there was also a land component to the trade. Canadian vehicles often pulled right up to border crossings to transfer their loads over to American cars and trucks (often in sight of customs buildings where the officers apparently looked the other way) but the sea-based operations proved the most rewarding.

As Victoria's *Daily Colonist* noted in a short article in April 1922, word had it from Bellingham-based officers of the law that "the islands are bartering grounds for 'liquor runners' from British Columbia and this country ... The San Juan Archipelago offers an ideal field for operations." It would appear that there were a few Canadians bold enough to venture across the line in order to garner a better return. As Washington state sheriff Al Callahan informed the reporter, "... Sucia Island [a small San Juan Island lying west of Bellingham, close to the border], the prettiest of them all, perhaps, is the clearing grounds for cargoes." But Callahan did note that this was only of late. Heretofore, he explained, the American boats had to go into British waters to get their cargos, but now the British runners from Victoria and Vancouver were delivering up their goods on the American side and charging for their risk. (Even though most of the crew probably lived on BC's West Coast at the time, most them were of English and Scottish background. Once the Dominion of

Canada's first transcontinental railway, the Canadian Pacific Railway, reached the nation's Pacific coast in 1886, English and Scottish immigrants were soon flooding the province of British Columbia. Of course, the Americans were quick to identify them all as Britishers because of their accents.)

As it was, the newspaper continued, "Tales of crime often trickle in from the exchange grounds. 'Knock overs' are said to be frequent. Freebooters range the waters and take possession of the cargoes, not caring whether their victims are Britishers or Yankees. There is a none too friendly spirit between the rum runners."

As it stood, Canadian rum runners were well within their rights

OASES OF SCOTCH
. IN DESERT OF SEA

KEEN DEMAND FOR ISLANDS IN GULF OF GEORGIA

Used as Bases for Liquor Running Enterprises, According to Seattle Newspaper

"Islands in the Gulf of Georgia have suddenly become in keen demand," says a Vancouver dispatch to The Seattle Post-Intelligencer yesterday.

"Mr. A. E. Craddock, of the County of Limerick, Ireland," the dispatch continues, "has purchased the north end of Prevost Island, half-way between Vancouver and Victoria.

"Two other islands recently changed hands, one going to a prominent British distiller, and the other to a San Francisco capitalist, who has money invested in Pacific Coast liquor running ventures.

"Of several purchases recently made in the Gulf, only one is credited as being other than for a rum-running depot.

"Publicity given Smugglers' Cove, Discovery Island, recently, has driven rum-runners to other islands in the Gulf. More than twenty of these little dots of land are credited with being each a Scotch oasis in a desert of salt water."

An article from the Victoria *Daily Colonist*, Marine & Transportation page, April 26, 1924. BC Archives newspaper microfilm files.

and not breaking any laws, as long as they stayed on the Canadian side of the border where they also felt relatively safe from hijacking. Most BC boat operators who took up rum running considered themselves just ordinary businessmen who were providing a delivery service by passing over their cargos to American vessels well within Canadian waters, preferably at any number of the convenient, out-of-the-way bays and coves hidden throughout the small islands on the Canadian side of Haro Strait. These included Sidney, D'Arcy and Gooch Islands and East Point on Saturna Island. The Discovery and Chatham Islands—right across from Oak Bay's popular Willows Beach—proved exceptionally ideal since they were only minutes away from US waters with a fast boat.

Philip Metcalfe explained it all well in *Whispering Wires: The Tragic Tale of an American Bootlegger*, his biography of Seattle bootleg kingpin Roy Olmstead. Numerous Canadian boats, some as small as 18 feet, which he claimed were capable of packing seventy-five to a hundred cases, were working for the big liquor exporters and delivering up their cargo f.o.s. (flat on the sand) to American boats out behind Discovery, Saturna or D'Arcy Islands. The trip out and back took only a few hours and an enterprising boat owner could easily make five or six trips a month and collect a profit of a thousand dollars (around fourteen thousand dollars today). In 1922, the Canadian government granted more than four hundred "deep sea" clearances to small boats engaged in the liquor trade. This fleet of fishboats, packers and tugs grew by the week and was soon recognized as the "whiskito fleet." Even with these fabulous returns, some more adventurous types were still willing to take the chance and run their load across the line into American waters to do even better.

In April 1924, Victoria's *Daily Colonist* noted that properties on a number of the islands in the busy shipping channel were in hot demand. "Mr. A. E. Craddock, of the County of Limerick, Ireland, has purchased the north end of Prevost Island" and "two other islands recently changed hands, one going to a prominent British distiller, and the other to a San Francisco capitalist, who has money invested in Pacific Coast liquor running ventures ... Publicity given Smugglers' Cove, Discovery Island [a local name] recently, has driven rum-runners to other islands in the Gulf. More than twenty of these little dots of land are credited with each being a Scotch oasis in a desert of salt water."

One very convenient transfer point, just over four miles directly east from the village of Sidney on Vancouver Island and lying next to Gooch Island, is Rum Island, which happens to be three-quarters of a mile from the US–Canada boundary out in Haro Strait. Smugglers Nook, historically another favoured rendezvous and stash point, whether the contraband was wool from the local sheep ranches, opium or booze, was located off the southeast end of North Pender Island.

With any contraband substance, the more risk taken importing or exporting a product, the more rewarding the financial return will prove if the undertaking is pulled off successfully. The export of alcohol across the line during US Prohibition was no exception. While a quart of good Scotch only cost $3.50 in Seattle prior to Prohibition, once the Volstead Act went into effect and bootleggers were still scarce, the price soared as high as $25 a quart. By 1924, it was selling curbside from anywhere from $6 to $12 a bottle. In the early years, American rum runners were paying anywhere from $25 to $40 for a twelve-bottle case (over $300 a case in today's currency) depending on the particular brand of Scotch, bourbon or gin delivered up by Canadian boats, which could be sold for up to $70 a case in Seattle.

The US government, of course, was annoyed when a flood of booze began pouring across the line—whether by land or sea—and put pressure on their Canadian counterparts to impose tougher laws on rum running. Still, as far as the Dominion government was concerned, it wasn't Canada's responsibility to enforce America's liquor laws since the nation's big export houses were private concerns that were licensed to ship liquor anywhere in the world. Still, on the surface, the federal government in Ottawa appeared to be somewhat willing to preserve good relations with its neighbour to the south, but it was most reluctant to forego the tax revenue generated by this particularly lucrative export trade. As far as Canadian government officials were concerned, everything was perfectly legal as long as all the customs and clearance paperwork was filled out properly. But in order to pacify their counterparts in Washington, DC, and at the same time boost government tax revenues, in the first year of the trade's operation, Ottawa raised the fee for a liquor export licence from three thousand dollars to ten thousand dollars.

In response to the federal government placing this steep tax on all liquor export agents, and with Prohibition firmly entrenched south of the border, it didn't take long for a number of clever liquor merchants to figure out that it might be better if they all banded together. This way the group only had to pay one ten-thousand-dollar licence fee for the whole lot of them.

As a result, Consolidated Exporters Corporation Limited came into being on August 25, 1922, with offices at 1050 Hamilton Street in downtown Vancouver. The stated object of the new company was to serve "as wholesale, import and export merchants, dealing with all classes of goods, merchandise, and wares and to buy, sell, prepare, market, handle, import, export, and deal in wines and alcoholic and non-alcoholic beverages of all kinds whatsoever insofar as the law allows the same to be done."

Among the more prominent shareholders listed among Consolidated Exporters' twenty-eight directors in October 1923 was Samuel Bronfman, who had already made a name for himself operating hotels and liquor outlets in Canada's prairie provinces prior to Prohibition. The Bronfman family started up their own distillery in Montreal in 1925 under the corporate banner of Distillers Corporation Limited and later acquired Joseph E. Seagram & Sons, whose Ville La Salle plant soon grew to be one of the world's largest distilleries. Another individual who was to really benefit from joining the consortium was Henry Reifel who, along with brothers Jack and Conrad, owned Vancouver Breweries Limited. They later acquired BC Distillery up the Fraser River in New Westminster in 1924. When he was interviewed by Imbert Orchard, who taped hundreds of BC pioneers for CBC Radio in the 1960s, Vancouver lawyer Angelo Branca said that the "industry was all corralled by the Reifel company. They made millions and millions of dollars." Although Henry Reifel was appointed one of the seventeen directors upon incorporation and also served as the secretary of Consolidated in the early years, the Reifel interests set up their own separate

CONSOLIDATED EXPORTERS
THE GIANT CONGLOMERATE

According to the Consolidated Exporters memorandum of association, the companies who put their heads together and federated their interests to create one of the largest liquor export conglomerates during Prohibition were:

- National Exporters Limited
- David Liquor Company, Limited
- Pither & Leiser Limited
- Rithet and Company
- Calgary Exporters, Limited
- Nat Bell Wine Company Limited
- Gold Bond Company Limited
- B.C. Liquor Company Limited
- Consumers Export Company Limited
- Brotman's Limited
- Lloyd and Son Limited
- Dominion Liquor Company Limited
- Glasgow Traders Limited
- J. G. Brooks Limited
- Dominion Trading Company Limited

Consolidated consisted mainly of British Columbia brewers, distillers and hotel-owning liquor agents. Along with the Vancouver directors and Olier Besner up the coast in Prince Rupert, there were also some interior-based companies, several of whose shareholders were also directors of Consolidated Exporters. These included British Columbia Exporters Limited; Fernie Exporters Limited; David Zimmerman, merchant; and the Strathcona Export Liquor Company Limited, which were all based in Fernie, along with the Globe Liquor Company Limited in Grand Forks.

The Consolidated Exporters Corporation Limited Warehouse, 1050 Hamilton Street, downtown Vancouver, BC, sometime in the early 1920s. Stuart Thomson fonds, AM1535 CVA 99-3487, City of Vancouver Archives.

Although not listed on its 1922 incorporation memorandum, the major Vancouver supplier affiliated with Consolidated Exporters was United Distillers Limited. The company was founded in Baltimore sometime before Prohibition, and its Vancouver directors were Isidore "Jack" Klein, Albert L. "Big Al" McLennan and George William Norgan, who also all happened to be on the board of directors of Consolidated Exporters as well. United Distillers' plant was located in Vancouver's Marpole neighbourhood and was strategically positioned facing the Fraser River. The export business itself was run out of the Credit Foncier Building in downtown Vancouver.

shipping company to handle the liquor produced by their British Columbia–based breweries and distilleries.

In order to carry on the business of wholesale and retail sales of liquor, it was noted in the Consolidated Exporters memorandum of association that they were able to "produce merchants, commission agents, manufacturers' agents, brokers, importers, exporters, ship-owners, charterers, charterers of ships, and other vessels." As such, this arrangement soon proved its worth, especially its sea-going arm, and by the end of the Prohibition years in late 1933, the various shareholding companies had done exceedingly well.

As Philip Metcalfe so aptly pointed out in *Whispering Wires*, it wasn't long before, "like a fabulous sea monster, its head stuffed with American greenbacks, Consolidated had tentacles reaching down the coast to Mexico and Colombia, across the Isthmus of Panama to Belize and eastward all the way to the island of Cuba. In the final years of rumrunning, the company established depots in Tahiti ... and China."

Throughout the Prohibition years, a steady stream of liquor consignments arrived into Vancouver harbour from Glasgow, London and Antwerp aboard Canadian Government Merchant Marine Ltd. freighters, Harrison Direct liners and Royal Mail steam packet ships. The liquor was brought right into Ballantyne Pier in Burrard Inlet, where it was put into government security. Here the finest brands of Scotch, rye, champagne and brandy were stored in bond. (Dominion Bureau of Statistics figures revealed that between 1925 and 1929 the value of all liquor imports into Canada grew from $19,123,627 to $48,844,111.) From this bonded warehouse, case upon case of liquor was loaded quite openly aboard "mother ships," the steamships or big ocean-going lumber schooners that sailed south to sit off the US coast in international waters out on "Rum Row." All that was required of the owners of the cargo was that they post a bond and swear on the vessel's clearance papers that it was destined for a Mexican, Central or South American port.

Stephen Schneider, associate professor in the Department of Sociology and Criminology at Saint Mary's University in Halifax, pointed out in *Iced: The Story of Organized Crime in Canada* that the underground liquor trade was driven by the most basic of economic laws: that of supply and demand. He also notes that "the Canadian conglomerates that grew fat off Prohibition were some of the first corporations to be vertically integrated, handling all aspects of their trade including production, distribution, sales, export, financing and marketing. They were also the most corrupt, unethical, and duplicitous corporations to ever operate in this country ... and did not shy away from selling to dangerous criminal syndicates ... They forged export documents, fraudulently listed the consigned destination of exported liquor as anywhere but the US, used counterfeit landing certificates from foreign ports, set up shell front companies, misrepresented the contents of their products, forged liquor labels, and bribed customs officials." As explained by Captain Charles H. Hudson, a decorated World War I Royal Navy veteran and marine superintendent manager, or "shore captain," for Consolidated Exporters in its later years: you could clear for any port in the world, but you didn't actually have to go there.

Hudson told Ron Burton how it all worked in an interview. "So when we wanted to load, we went alongside [Ballantyne Pier] ... and loaded quite openly. We had to put a big deposit down, I think it was twenty dollars a case. Say we cleared for La Libertad [Mexico] ... We'd send our agent down there and he'd pay somebody off a few hundred bucks and he'd send back a release [the landing certificate] ... and we presented this to customs ... and all our money was returned!" Then it was only a matter of getting the clearance papers back out to the mother ships sitting out on Rum Row.

In February 1924, with liquor pouring across the line, Ottawa finally caved in to the constant pressure from Washington, DC, but only resolved the issue with a half-hearted measure. Now, it announced that it was no longer going to allow the smaller vessels in the rum fleet to clear for distant points such as Mexico or Central America, which the owners quite obviously never intended to reach. To get around this, liquor merchants and export companies changed their clearance papers for destinations along the BC coast.

In the end, these measures didn't slow things down much at all. From then on, it was simply a matter of consigning several hundred cases of whisky to an up-coast port like Bella Coola or even Bowen Island, regardless of the fact that there was hardly anyone living out there to drink them. Meanwhile, boat and cargo steamed off in the opposite direction towards Puget Sound or farther down the American coast. "I'm sure that if the records for that period were examined, places like Bowen Island would show a per capita alcohol consumption that far exceeded human capability!" This according to Johnny Schnarr, one of the most successful West Coast rum runners, as told to his biographers, niece Marion Parker and Robert Tyrrell, in *Rumrunner: The Life and Times of Johnny Schnarr*.

A few months later, Washington, DC, itself made a move to curtail the ever-growing flood of liquor pouring into the country. They signed a revised liquor treaty with Great Britain which was proclaimed by President Calvin Coolidge on May 22, 1924. At that time, Great Britain still represented Canadian interests and maintained jurisdiction and control over Canada's territorial waters.

The revised treaty extended the standard three-nautical-mile international territorial limit to twelve nautical miles or one hour's sailing from the US coast. This was in order to make it harder for the smaller and less seaworthy craft to make the run from international waters to the beach. But most importantly, the treaty allowed for any ship sailing under the British flag and suspected of carrying liquor to be searched if it was within an hour's steaming distance of the US coast. As it was, small, fast boats were easily outrunning Coast Guard ships to dock or land in small, out-of-the-way bays and coves or even run up a river along the California coastline where the cargo was landed and transferred to waiting automobiles or trucks.

Of course, federal officials in Washington, DC, continued to pressure their counterparts in Ottawa to restrict companies from loading liquor in Canadian ports, only to deliver it up to American vessels. Officials were well aware of the fact that they never did stop in at any Mexican, Central or South American port to unload as declared on their fake landing certificates. It wasn't until 1930, however, that the Canadian government introduced a bill to amend the Canada Export Act requiring that all Canadian ships clearing for a foreign port loaded with liquor proceed to that port and actually discharge their cargo there. Although it was not officially passed until March of that year, rum running interests like Consolidated Exporters had already found a clever way to legally circumvent this requirement by actually delivering liquor to their stated destination. It was simply a matter of locating a friendly transshipment port with suitable warehouse facilities and, most of all, easygoing port authorities who weren't really concerned about whether a particular cargo was actually unloaded in their port or not.

Tahiti became the destination of choice, and mother ships or their supply ships settled into a routine of sailing from Vancouver to the tropical customs-free port of Papeete, much to the delight of their crews, before crossing back over the Pacific to Mexico to continue on with their trade (in the latter years of Prohibition, Rum Row was finally situated just south of Ensenada, Mexico). The whole undertaking proved straightforward and easy to manage. Once the Canadian merchandise arrived in Tahiti, all it required was a little creative paperwork, and most likely paying off port and customs officials, and sailing back across the Pacific to Rum Row. Then once the mother ship was sitting out on Rum Row and its cargo had been delivered up, stocks were then replenished from Tahiti aboard sailing schooners ranging from the two-masted *Aratapu* (Peruvian) to the big three-masted *Marechal Foch* (Tahitian), as well as several others. The rum fleet was also kept well supplied from other customs-free ports besides Tahiti.

Regardless of Rum Row's location or the logistics, the whole operation remained above-board, involving nothing criminal or untoward. As Hudson stressed in his interview, "We operated perfectly legally. We considered ourselves philanthropists! We supplied good liquor to poor thirsty Americans who were poisoning themselves with rotten moonshine ... and brought prosperity back to the harbour of Vancouver ... Many people thought of rum running as a piratical trade ..." Instead, "it wasn't anything dangerous. Simple, clean business operation from start to finish, no firing or hijacking, nobody lost or drowned or killed, good wages, paid well, a bonanza in Vancouver because Vancouver was in the dumps!" This proved especially true during the early years of Prohibition, when rum running remained mostly a trouble-free way to earn a fast dollar, especially for those who might own an old fishboat, packer or towboat with a reasonably sized hold, if not crewing aboard a mother ship. The only real threat was from hijacking. It was the Americans who took all the risk and who needed the fast hulls in order to outrun their own Coast Guard once back in US waters. Even so, the US Coast Guard was dreadfully lacking in patrol boats at the onset of Prohibition.

Still, there were also some major costs to bear that played out over the fourteen years that the Volstead Act remained in effect south of the line. Of course, throughout the United States there were particularly tragic consequences that arose from Prohibition. Right from the very start, bootlegging soon fell under the command and control of America's very own sophisticated management entities, those of the criminal element. And as the 1920s progressed, the downside to all the high adventure and romance of running booze out of Canadian waters began to reveal itself. Even so, the tales of high adventure around local waters and down the American coast still resonate, especially following the publication of memoirs later in life by some of the characters that took advantage of Prohibition in order to earn a good dollar.

Chapter Two

SHOWDOWNS IN THE STRAIT

While rum running was generally a relatively harmless and violence-free undertaking along the West Coast, especially if carried out in Canadian waters where literally hundreds of transactions took place, there were some dramatic events involving hijacking that grabbed the headlines in the early years.

"Pirates Prey on Rum Runners" is the headline citizens of Victoria were shocked to read upon opening their *Daily Times* evening newspaper on January 16, 1922. It was a fascinating albeit short tale, highlighting the dangers, especially to small operators, of making surreptitious liquor smuggling runs across the border. Apparently, Charles Boyes had just returned to Vancouver minus his fast motorboat and cargo of whisky, as well as a substantial bankroll. He had been running comparatively small cargos into Puget Sound points and on this particular trip, was carrying fifty cases of liquor.

Boyes informed a reporter that he had been playing a "lone hand" but had decided the week before to take a partner and so invited "Omaha Whitey," who was over in Victoria at the time, to join the enterprise. Upon his arrival in Vancouver, Whitey learned that Boyes was overdue from his trip. Then he received a wire from Bellingham asking him to forward funds as Boyes was in some sort of trouble and very eager to get back to Vancouver.

After he had crossed the line into Washington state waters and was cautiously looking for the agreed-upon landing place with his running lights doused, Boyes was overhauled by a larger and faster boat, also running with no lights. When the unidentified boat ran up beside him, three men jumped aboard with revolvers in hand, shouting, "You are under arrest. We are US federal officers." They quickly set upon him to carry out a body search and relieved him of his $2,100 emergency fund, saying they were seizing the cash as evidence. (Since a 1922 US dollar would be worth close to $14.50 in 2017 dollars, Boyes was carrying around thirty grand in today's money.) After leaving two men in charge of his boat, they forced Boyes aboard their craft and set a course for Bellingham.

Before they got there, however, they landed on a quiet beach where they shoved Boyes into an automobile. They informed him he was on his way to jail to await trial in Bellingham. But upon reaching the outskirts of the city, the car pulled over. Here the "officials" ordered Boyes out of the car and told him to beat it.

As he stood there and watched the car race away into the darkness, he quickly realized that he had been the victim of pirates. Not willing to admit he'd lost everything, Boyes made "careful inquiries" around Puget Sound to try and locate his missing boat, fully aware of the fact he couldn't appeal for aid from American officials. Once he resigned himself to the fact that he was out of luck and had lost it all—the boat, his cargo of liquor and his bankroll—he wired for money and returned to Vancouver. Still, Boyes had got off lucky since hijacking, especially in American waters, only got more violent over the next few years as Prohibition set in. As it happened, the manufacture, distribution and imbibing of alcohol in the United States of America had been illegal almost two years to the day at the time of Boyes's loss.

The most feared of the American hijackers were the Egger brothers, Theodore "Ted," thirty-two years of age; twenty-nine-year-old Ariel "Happy"; and Milo "Mickey," who was twenty-six; famous for their black-hulled speedster *Alice*, powered by a two-hundred-horsepower Van Blerck gasoline engine. But it wasn't just their fast boat that made their reputation in the early 1920s. Rum runner Hugh "Red" Garling described them as the most unscrupulous, brutal, hardened criminals known throughout the Pacific Northwest during their brief career. He noted that for eighteen months between 1922 and 1924 the Egger boys brought to the Gulf Islands, San Juan Islands and Puget Sound waters "an interval of terror." The returns for American rum runners were particularly rewarding since they were able to buy whisky from Canadian carriers for around $25 a case and then sell it in Seattle for up to $225 a case. But the Eggers decided this wasn't quite good enough and turned to piracy in order to obtain it gratis.

One day while tied up at the float at Discovery Island, Johnny Schnarr, a Canadian who was earning a good living "jumping the line," and his partner noticed a boat painted completely black pulling into the bay. He described the two men as rough-looking characters, both hawk-faced and raw-boned, and he didn't like the looks of them at all. When they began asking all sorts of questions about the price of liquor but didn't seem all that interested in buying any, Schnarr and his buddy got suspicious. *Alice* shadowed them for a while after Schnarr and his partner picked up their liquor and set off for Anacortes, but the evil-looking boat finally turned away and headed off in another direction looking for easier pickings. Schnarr figured his demonstration of skill shooting bottles out of the water with his Luger while waiting at the Discovery Island float prob- ably discouraged the desperate-looking characters from bothering with them. It was only when they were back on Discovery Island a week later that they learned that they'd met up with a couple of the notorious Egger boys.

Hugh Garling claimed that the first victim of the Eggers' cross-border piracy was Tom Avery of Vancouver in his boat *Pauline*. The Eggers relieved him of the 128 cases valued at $5,376 he was hauling for Great Western Wine Company.

The boys' next attempt at hijacking was the 73-foot fish packer *Emma H*, which they came across anchored in Smuggler's Cove on the inside waters of the Discovery and Chatham Islands in May

Emma H, a two-masted halibut schooner 73 feet in length, was soon to be recognized as "the Hulk" and put to use in a more lucrative trade—transferring liquor orders to Yankee boats out in Haro Strait. From the collection of Canadian Fishing Company.

1923. The Eggers rowed over in a dinghy with automatics drawn only to be met by Captain Emery and crew armed and ready for them. Somehow, after convincing Emery that their boat was too cold and damp, they managed to talk themselves into spending the night aboard *Emma H*. A wary Emery agreed but posted an armed guard to keep an eye on them throughout the night. Giving up on this particular knockoff, since the Eggers were probably reluctant to shoot it out, the trio returned to *Alice* and headed off for nearby D'Arcy Island. Here they found another fish packer being put to good use in the trade, the 31-foot *Erskine*, whose cargo they were more successful in purloining at gunpoint. They ordered its captain, a fellow named Steele, to proceed at full speed across into Washington state waters to Dungeness Head, just east of Port Angeles on the Olympic Peninsula, where they beached *Erskine*, disabled her engine, cached half her cargo of liquor, transferred the other half to *Alice* and took off. Unfortunately for the Eggers, they'd chosen the wrong boat to hijack since it was running liquor for Roy Olmstead.

Olmstead was a former Seattle police officer who displayed a natural talent for organization and administration when he put together a sophisticated wholesale rum running operation in Washington state. After many Seattle residents became tired of the vicious competition and chaos in the illegal liquor market in the early years of Prohibition, Olmstead encouraged some prominent local citizens to invest in an international company to operate and control a wholesale liquor enterprise, thereby bringing some sanity to the risky and illegal business. Olmstead got off to a good start by first bribing both American and Canadian customs officials so he could lay out a safe rum running route between the two countries. The first exchange point he established was D'Arcy Island just outside of Victoria, where the Eggers chose to hijack *Erskine* and make off with its shipment. This infuriated Olmstead. The Eggers messing with the wrong guy, one who had all the right connections, was probably a big factor that led to the American authorities finally catching up with them a year later. But, for whatever reason, the attempted holdup of *Emma H* and hijacking of *Erskine* just off the Victoria waterfront didn't make the news in local BC papers. Perhaps the victims weren't all that willing to share details of their unsavory operations with a reporter. (While both Johnny Schnarr and Hugh Garling gave very detailed accounts of some of the rather nasty incidents that went down in local waters years later, they always remained discreet and didn't give away the names of their sources even though the incidents they recounted had occurred more than fifty years earlier.)

The next known at-sea stickup perpetrated by the Eggers was that of the Canadian-owned *Lillums*, near Hope Bay on North Pender Island in August 1923, when the Egger gang made off with sixty-three cases of liquor. Adolf Ongstad and Jack Webstad were operating the runner. Webstad had gone off to look for a guy named Thompson who they were to rendezvous with at Bedwell Harbour on South Pender Island, leaving Ongstad to stay with the boat. Sometime later, while dozing in his bunk, Ongstad was disturbed by the sound of a powerful engine close by, and when he peered through the porthole, saw *Alice* coming alongside. Then the cabin door was pushed open and Ongstad found himself looking down the barrels of pistols in the hands of the Egger brothers and their accomplices, ex–prize fighter P. K. Kelley and two thugs named Pfleuger and "Tiny" Palmer. Once aboard, the lines to the dock were cut, *Lillums* was secured alongside *Alice* and they headed out into Navy Channel where the liquor cargo was transferred over and *Alice* raced off into American waters.

It wasn't until March 1924 that the gang was back in the Gulf Islands again. *Kayak*—manned by two Seattle runners, "Feathers" Martin and engineer Joe Edwards—was patiently waiting to rendezvous with the Canadian boat *Hadsel*, operated by Fred Davidson and his engineer, Adolf Ongstad, at Peter Cove at the south end of North Pender Island. *Kayak* was to take on an order of 293 cases of Scotch, which *Hadsel* had loaded in Vancouver and which had already been paid for. That night, while *Kayak* sat moored awaiting the arrival of the BC rum runners, four masked and armed men in a dinghy snuck aboard the boat and caught the two Seattle men by surprise. After Martin and Edwards were tied up, the visitors settled in for the night frying bacon and eggs, brewing coffee and generally making themselves comfortable. When *Hadsel* arrived just after daybreak the next morning, Mickey Egger rammed a .45 automatic in Martin's back and told him to stick his head out of the hatchway and signal across that everything was all right and to come alongside. Once *Hadsel* was within hailing distance, they jumped up shouting, "Stick 'em up!" and started shooting, riddling the wheelhouse with seventeen shots. Davidson automatically reached for his Winchester but after taking a couple of shots in the shoulder and ankle, quickly surrendered and the Egger gang jumped aboard. The four desperados unceremoniously dumped the two Canadians onto *Kayak* and took off in *Hadsel*. The Eggers then ran her into a nearby bay where they transferred the liquor cargo over to their boat, *Alice*. Once loaded, they opened up the throttle to roar back into American waters.

San Francisco police finally arrested both Happy and Mickey on a charge of highway robbery in November 1924, while Ted was picked up sometime later in 1925 in Tacoma, and the Eggers' reign of terror, both in Puget Sound and Canadian waters, came to an end.

But, as Johnny Schnarr pointed out, by that time the threat of violence had one side effect: all those involved in the business started packing firearms, while some were scared off completely. He said that it was getting to the point where there were just too many trying to make a fast buck running liquor in local waters and the resulting competition was bound to drive down the delivery price. Regardless of the danger posed by the likes of the Egger boys, Canadians involved in the liquor trade still remembered the early years of us Prohibition as golden ones.

Around the same time that the notorious Eggers were being hunted down in the States, two Canadians, William J. Gillis and his seventeen-year-old son, fell victim to the brutal hijacking of their 53-foot wood packer *Beryl G* off Sidney Island by two hard-bitten thugs. This time the rum runners not only lost all their liquor but their lives as well. Don Munday, who wrote up the tragic tale in the BC Provincial Police magazine, *The Shoulder Strap*, in 1940, called it "the most ghastly episode of lawlessness in British Columbia

Once word was out around local waters about the brutal hijacking of the *Beryl G*, the local rum running fraternity vowed to hunt down those responsible for the murder of Captain Gillis and his son and shoot them on sight. Leonard McCann Archives, LM2018.999.032, Vancouver Maritime Museum.

waters during the rum-running years." He also noted that it resulted in a "relentless police effort to solve a cold-blooded killing carried out with diabolic cunning."

On September 17, 1924, Chris C. Waters, light keeper at Turn Point, Stuart Island, on the US side of Haro Strait, noticed a boat drifting northward past the lighthouse and figured it must have had engine troubles and broken down. Waters crossed the island to ask the husband of the postmistress to bring his gas boat around, and the two went out to see what the problem was. As they pulled alongside the drifting craft, which they now identified as *Beryl G*, they banged on her hull. But there was no response. Thinking they were dealing with an abandoned boat, they decided to put a line on and tow her into Stuart Island but when they came alongside they noticed bullet holes in the fo'c'sle door. Once aboard, they were shocked to discover bloodstains all over the hatch covers and along the deck, as though something heavy had been dragged over them. Up on the bow they came upon a pile of blood-soaked clothing, and inside there was even more blood all through the companionway and spread over the stove and the settee. Now fully aware that something terrible had gone down, they contacted the US Coast Guard, who quickly notified the British Columbia Provincial Police, as *Beryl G* was registered in Vancouver.

In a story on the incident in the October 1989 issue of *Harbour & Shipping*, Hugh Garling said that *Beryl G*, which was named after Captain Gillis's daughter, had been used for freighting supplies to logging and mining camps up the British Columbia coast prior to becoming a packer. Don Munday, still a well-known BC mountaineer and naturalist, worked up a story on the most violent hijacking of a vessel in BC waters, which appeared in the winter 1940 edition of *The Shoulder Strap*. Munday wrote that besides operating her as a freight boat, Gillis was also trying to make a go of it with a small oyster cannery, but a poor season had set him back. The subsequent investigation revealed that Gillis had been approached by an agent for well-known Seattle rum runner Pete Marinoff, who was looking for a Canadian to freight liquor for him.

Gillis accepted the offer to make a fast buck and was sitting on the anchor around midnight on the east side of Sidney Island after making his fourth transfer of 110 cases from *Beryl G*'s cargo of 350 cases across to Pete Marinoff's boat, *M-453*. (Marinoff had paid twenty-eight dollars a case.) *M-453* was an ideal rum runner, a twin-screw "fast launch," 56 feet long, powered by twin three-hundred-horsepower Sterling engines and capable of thirty knots. Conflicting reports say that *Beryl G* was carrying anywhere from 350 to 600 cases of liquor at the time, all loaded from "the rusty old British freighter *Comet*" that was sitting outside both American and Canadian territorial waters off Vancouver Island's west coast.

Marinoff sent the money out for his order of 110 cases but no money was found aboard the abandoned vessel when provincial police examined the ill-fated craft.

But all they came across were the bloodstains and bullet holes in the woodwork, along with what appeared to be signs of a struggle: all evidence suggesting that murder had been committed in the midst of a hijacking. While police authorities on both sides of the border set out to locate and capture those responsible, the local rum running fraternity also became involved. They were outraged that their lucrative trade, which was sustaining so many mariners on both sides of the border, was now being threatened by the criminal element. Victoria's *Daily Colonist* reported on October 10, 1924, that in Vancouver, leading rum runners had banded together to run down the murderers of Captain Gillis and his son and vowed to shoot them on sight. Still, it was British Columbia's very own Provincial Police force that would track down the culprits.

Vancouver Island fisherman Paul Strompkins ("a big, fat-faced Pole, a former Manitoba farmer turned fisherman," as Don Munday described him) was caught and arrested on Vancouver Island. He'd been picked up because of his past record and Canadian customs officers confirmed that he had been smuggling beer. Also, he happened to have introduced a customs officer in Beacon Hill Park to several Americans, one being an individual named Owen Baker. After a few other clues emerged that put him under suspicion, Strompkins broke under constant grilling and promised to come clean. After careful inquiries down along the Seattle waterfront, Inspector Forbes Cruickshank, in charge of the Criminal Investigation Branch and recently assistant commissioner of the BC Police, went with two King County detectives and arrested Charley Morris. Then, after a reward of two thousand dollars was offered for each man, Owen Baker was picked up in New York, where he was working on a barge under the alias of George Nolan, while Harry Sowash was picked up in New Orleans by local detectives who recognized his photo in a magazine circulated among US police forces. All three were then extradited to British Columbia to stand trial.

At the trial, it was revealed that Strompkins had embarked the three American desperados in Cadboro Bay close to the Royal Victoria Yacht Club in his "beer boat," *Denman No. 2*, to take them out to Sidney Island. (The trio probably avoided using *Dolphin*, the boat they'd chartered from Albert Clausen, who ran an auto repair shop in Seattle, since she was too easily recognizable as a fast hull probably used for running liquor.)

Cecil "Nobby" Clark, who served for thirty-five years in the BC Provincial Police and became the service's unofficial historian, writing numerous articles for local papers along with a series of books on BC crime, described the three ruthless characters in one of his stories, "Hijack Route to the Hangman": "Owen W. Baker was tall and gangly, with an Adam's apple that moved convulsively in his scrawny throat. A lock of hair falling over his forehead gave him a folksy look that belied the savagery he later revealed. Harry "Si" Sowash was somewhat younger, and with his crewcut, broad shoulders and rugged build, he could have been taken for a university football player. In keeping with this academic impression, he read a lot and knew his way around in Greek and Roman history." The two met in McNeil Island penitentiary down in Puget Sound. Owen "Cannonball" Baker was doing a five-year stretch for white slavery while Sowash was serving two for selling stolen airplane parts. They were both released during the heyday of rum running and were soon taking advantage of the opportunities it presented.

In early September 1924, Baker, Sowash and their new accomplice, Charley Morris, had been cruising off the west coast of Vancouver Island in *Dolphin* for a number of days, searching for liquor caches, when they happened to notice the *Beryl G* off Sooke Harbour, headed for Victoria waters. The trio soon determined that the boat was shuttling booze to the east side of Sidney Island, where her cargo was delivered up to American boats. Once Strompkins

Owen "Cannonball" Baker was described by Don Munday, who worked up a detailed story on the hijacking of the *Beryl G* in 1940, as "tall, dark, glib-tongued with a pasty complexion." He said Baker also had a wife and child, whom he apparently wasn't supporting all that well. Royal BC Museum and Archives, Image B-05758.

turned king's evidence, it didn't take long into the trial to learn the gruesome details of the murders. After dispatching father and son, the three American thugs handcuffed the two bodies together, stripped them of most of their clothes and then ripped the bodies open with a butcher's knife so they wouldn't float once heaved overboard. Found guilty of murder, Baker and Sowash were sentenced to hang on September 4, 1925, while Morris received a life sentence. Strompkins was also charged with murder, but since the Crown was unable to find any evidence directly linking him to the murders, he was discharged. The date of execution for Sowash and Baker was advanced and on January 14, 1926, they were both hanged in BC's Oakalla prison.

One wonders if perhaps it might have been Cannonball Baker who hijacked Charles Boyes back in January 1922 and made off with his boat, liquor and bankroll. (Harry Sowash wasn't let out of prison until just a little before the *Beryl G* incident.) The modus operandi appears strikingly similar. According to Cecil Clark,

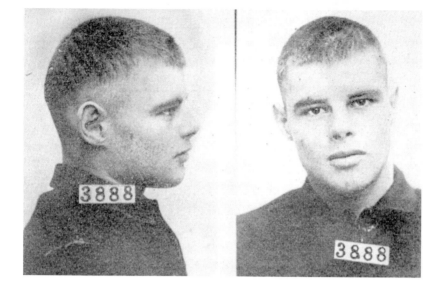

In the day before his execution on January 14, 1926, at Oakalla prison, Harry Sowash wrote in a letter: "A raging fire of regret, of remorse, a feeling of unfitness, a sense that I have cheated myself, torments me ... A short season of moral laxity is but the beginning of a swift, dizzy ride ... with a weary, stone path back." Royal BC Museum and Archives, Image B-05759.

Baker typically outfitted himself with a yachtsman's peaked cap, a blue blazer with double rows of brass buttons, flashlight and a phony police badge. Out patrolling local Puget Sound beaches in the dead of night, he and his cohorts would sneak up and surprise the "rummies" unloading their boats by jumping up shouting, "United States customs! Stay where you are!" After the surprised rum runners took off for their vehicles, Baker "confiscated" their booze.

The *Beryl G* murders put the whole rum running trade on edge. As a consequence, the generally peaceful enterprise started to take a nasty turn for the worse. All those involved, whether directly taking part or trying to put a stop to it, started keeping a rifle or pistol well within easy reach. Now, shootouts and gunplay started becoming commonplace in southern British Columbia waters and even off the beaches inside the "Tweed Curtain" of Victoria's plummy Oak Bay neighbourhood.

While accounts of violent encounters and hijackings by the criminal element out among the rum running fleet were grabbing headlines, US enforcement agencies weren't shy about resorting to gunplay in order to try and put a halt to the trade in Washington state waters. Meanwhile, their counterparts in BC were also known to reach for their firearms in order to keep up the appearance that the liquor trade was all legal and above-board on their side of the line.

Chapter Three

US COAST GUARD GETS TOUGH

On July 1, 1924, Victoria's *Daily Colonist* ran a story under the sensational headline, "Big Whiskey Cargo Seized by Police," which provided insight into how Canadians were interacting with their American brethren in local waters. "When the police launch *Dorothy*, a 32-foot yacht hired by the local police for use as a patrol vessel, arrived at the entrance to 'Smuggler's Cove,' Chatham Island, on Thursday afternoon, its occupants were treated to a fast display of scattering." The police launch came upon a dozen boats in the immediate vicinity and all but two turned about and made off when the launch drew near. When *Dorothy* entered the cove, she discovered three launches moored to the wharf. Seeing that one craft, identified as *M-332*, which indicated American registry, sat low in the water, the police boarded the vessel and a cargo of sixty cases of whisky, valued at $3,500, was found in her hold while the crew was nowhere in sight. A smaller craft, a Canadian boat, was found to contain a full cargo of cased beer, but as her owners had possession of all the necessary papers properly drawn, she was not molested. However, *M-322* was taken into Cadboro Bay and handed over to customs officers.

The article continued that on returning to the Chatham Islands later that day, *Dorothy* was headed for the third vessel, a rakish-appearing craft which, from her depth in the water, probably carried a good load. "When the police boat was some distance away, one of the members of the moored boat's crew cut the mooring rope, and despite threats to shoot from one of the officers on the *Dorothy*, the craft swung out to sea and in a remarkably short space of time had reached the horizon."

Discovery and the Chatham Islands, with their quiet coves, narrow channels and close proximity to the US-Canada boundary, were an ideal rendezvous point for rum runners. Here a fellow and his wife even anchored their boat out behind Discovery Island at what became known as that island's Smuggler's Cove during the Prohibition years and sold beer to the Americans running across the line. Once sales began to pick up, they realized there was even better money to be made selling hard liquor. The couple then built a small lodge on one of the islands, where they offered good home-cooked meals to all those who were there waiting to pick up or deliver a load. They also built floats out into deep water so there was room to tie up right in front of the lodge.

With the rewards being so high, Canadian operators soon learned that they still had to be prepared to defend themselves since they were sitting ducks to the hardened American criminal element that wasn't going to let any border get in their way. Those involved in trying to put a stop to the rum running weren't prepared to admit its benefits. Only the year before, June 3, 1923, *The Daily Colonist* ran a front-page story under the headline, "Rum Runners Hard Pressed: Are Making Last Stand on the British Columbia–Washington Border." In an interview in Spokane, Roy C. Lyle, Washington state federal Prohibition director, declared that a "tremendous flood" of spurious liquor (probably American moonshine) was now being offered in Seattle because rum runners were finding it more difficult to bring in liquor from Canada and that bootleggers and rum runners were making their last stand on the British Columbia–Washington border as a result of other provinces having been driven out the liquor export business. Nonetheless, whether his office was showing results or not, he hoped that laws would be amended to permit Prohibition officers to use confiscated automobiles, boats and airplanes in enforcement since, he lamented, his men were hampered by lack of equipment. But American law enforcement did eventually come into their own and could lay claim to some dramatic seizures and arrests by late 1924. And these involved vessels in the supposedly legitimate shipping and export trade operating out in international waters. As it happened, Canadian authorities were also making the odd seizure themselves.

Right beside the story on how rum runners in American waters were now "hard pressed," the *Colonist* reported that, on this side of the line, law enforcement was also having some newsworthy successes. Following their efforts to break up an (unnamed) "rum-running gang which has been operating for months past with 'brazen effrontery' between Canadian and United States points in the Gulf, Provincial Police officers and customs authorities yesterday [Monday, June 4] made sensational seizures in the vicinity of Discovery Island near the spot where, on Saturday, they captured two Canadian launches, alleged rum runners, and two American craft." The paper stated that great credit was given to the efforts of Provincial Constable Wilkie, Customs Officers Bittancourt and Norris, and Dominion Constable Harvey, for breaking up the ring. On Monday morning the four enforcement agents headed out in their launch, *Ark*, to look over the fish packer *Emma H* (or *The Hulk*, as she was better known around local waters. This was the vessel that the Eggers had unsuccessfully tried to hijack). The *Emma H* had apparently transferred her cargo over to *Cleegone*, a Canadian launch, the previous Saturday. It was seized and brought into Victoria the next day, Sunday. Once the inspection of *Emma H* was finished, she was left out on the spot and Constable Wilkie and the other officers continued with their investigation of neighbouring waters throughout the Chatham Islands. In a sheltered bay on Discovery Island, they came upon two power boats, one the *Standard* from Anacortes, the other unnamed. Two men found asleep on the *Standard* and one on the other vessel were caught by surprise and the crewmen were detained and brought into Victoria. And it wasn't over yet.

The same four enforcement officers returned to Discovery Island in *Ark* later Monday afternoon. This time they came upon a vessel, *Syren*, passing booze over to two Yankee boats who, when they saw the Canadian authorities bearing down on them, made off towards the American side. And despite every effort to overhaul them, they made good their escape, with *Ark* having to call off the chase at the Canada-US boundary line. *Syren* also attempted to make a fast exit, but was overhauled and one American, a gentleman named Kerr, was arrested. Both *Standard* and the unnamed vessel from the morning capture were detained and brought into Victoria. *Syren* was brought into Victoria later that evening and tied up at the Causeway. While no liquor was found aboard *Standard*, she was seized, since she broken Canadian law by entering the country at a place other than a regular port of entry without the excuse of emergency to justify her action.

By late 1924 American enforcement had also achieved some successful seizures and arrests, although they were dreadfully lacking in vessels and manpower. In a Seattle news release in June 1924 it was reported that four men were seriously burned and three wounded by bullets when the coast guard cutter *Arcata*, out for rum smugglers, pierced the gasoline tank of their boat lying in Mutiny Bay, eighteen miles north of the city. Ignoring the fact that a quantity of Canadian liquor was found on the vessel and it was running without lights, one of the wounded protested that they were all just out on a hunting trip. He claimed that *Arcata* opened fire with her one-pound gun without warning and fired four or five shells, and that he was hit while standing at the wheel. This capture was quite a praiseworthy feat for the *Arcata* since she left much to be desired in her role. She was only 85 feet long, top heavy and unfit for the open ocean, and had a top speed of only twelve knots.

On Saturday afternoon of September 27, 1924, the crew of the 50-foot tug-turned—rum runner *Ironbark* were fired upon by

THE VICTORIA CONNECTION

A retired British Columbia police officer who became the force's historian, Cecil Clark said in an article that ran in *The Daily Colonist*'s Sunday supplement, the *Islander*, in February 1969, that he estimated that there were about a dozen firms in Victoria alone exporting liquor during us Prohibition. He mentioned that he had obtained some particularly insightful information courtesy of "a gentleman now resident behind the 'tweed curtain' (Victoria's upscale Oak Bay neighbourhood): Victoria's Rithet Consolidated Limited handled the ever-popular King George the Fourth Blended Scotch Whiskey brand, Pither & Leiser handled Teacher's Scotch whiskey, while Western Freighters' specialty, Granny Taylor's Whiskey and Peter Dawson scotch."

Cecil Clark also interviewed retired customs officer Joe Dakers who, in recalling the comings and goings of the export trade out at Victoria's outer docks, spoke of cargos "that beggar description. Like the time he checked out 70,000 cases. It came in by Furniss Withy [a British shipping company] and went right out again to California." He said the bonded warehouse out at Ogden Point held up to 100,000 cases when it was chock-a-block full.

Victoria-based customs officials as they attempted to board the vessel in Cadboro Bay to check her clearance papers. Victoria residents were astonished to read in their Monday *Victoria Daily Times* the sensational headline, "Escapes hail of bullets from customs men's guns." Apparently, the crew of the 57-foot tug *Superior* were working on a log boom in Cadboro Bay when they were startled by the sound of gunfire and bullets humming over their heads and promptly sought cover. Slipping and sliding over the logs of the boom, they headed as far as they could from the paths of the flying lead. As for *Ironbark*, she was registered to Coal Harbour Wharf and Trading Company of Vancouver and owned by Archibald "Archie" MacGillis, who had already established himself as a major player in West Coast rum running by that time.

R. Harrap, the steward at the Royal Victoria Yacht Club, was interviewed by the *Times* reporter about the attempted boarding. Harrap stated that about noon that Saturday, a party of men (who turned out to be customs officers) came down through the grounds of the club and quickly made their way onto the float. Assuming them to be visitors from another yacht club, he didn't bother asking any questions. But when the strangers shoved two boats into the water, he ran down in an attempt to stop them. "Those aboard the *Ironbark* supposed the strangers were hijackers," said Harrap. "The customs men, whom I did not recognize as such at the time, were armed and none of them were dressed in uniform. The *Ironbark* had legitimate clearance I believe … The *Ironbark* slipped her moorings eventually, and as the customs men got to within twenty feet of her, suddenly shot forward and twisted out into the bay, running clear of the shots of the officials," Harrap added. In an attempt to explain their drastic actions, customs officials told the *Times* reporter that they did the best with the equipment they had, and they only intended to look over the launch's clearance papers. And when she did not stop on their warnings, they opened up on her.

ARCHIE MACGILLIS
A FOUNDING FATHER
OF RUM RUNNING

Archie MacGillis was already well situated when Prohibition came into effect south of the line in the early 1920s. At the time, he was busy operating his towing outfit, the Vancouver Courtenay Transportation Company, located in Coal Harbour, not far from the entrance to Vancouver's Stanley Park. His company ran barges across the Strait of Georgia to Vancouver Island. It soon proved an ideal location for rum running, since it was not all that far away from the bonded warehouses in Burrard Inlet where choice brands of Scotch, rum and brandy were arriving from Europe. Not one to miss an opportunity, MacGillis was soon being recognized as a "founding father" of West Coast rum running. He eventually put together at least four liquor export companies: Arctic Fur Traders Exchange, Coal Harbour Wharf and Trading Company, the Canadian-Mexican Shipping Company and Iron Bark Exchange.

The first of the export companies was the Canadian-Mexican Shipping Company, incorporated in May 1921 by three local investors to take advantage of the opportunities offered during Prohibition. One year later, "Archibald MacGillis, Shipowner" entered the books as a company shareholder.

MacGillis's companies owned several large mother ships that operated off the US and Mexican coast as well as a number of smaller, fast launches for use closer to home. MacGillis caught on early that strong, fast boats were required if one was to succeed in the trade. In 1922, he picked up two retired US Navy subchasers (*SC 293* and *SC 310*, renamed *Etta Mac* and *Trucilla* respectively). The Coal Harbour entrepreneur was soon taking full advantage of the fast, narrow-hulled subchasers with their minimal housework and low profiles, which made them extremely difficult to spot out on the water, especially after they were painted grey or black. Soon both were "jumpin' the line."

Eager to jump into the sea-going liquor export trade in a more serious way, MacGillis purchased the three-masted auxiliary lumber schooner *Leonor* later that year and renamed her *Coal Harbour*. She was 127 feet in length and was launched as *Lottie Carson* in July 1881 from the Hall Brothers Shipyard in Port Blakely, Washington. The following spring, he bought a vessel with even more carrying capacity, the retired 247-foot, 6-inch five-masted lumber schooner *Malahat*, and that was just the start.

Ironbark was registered to Coal Harbour Wharf and Trading Company, which was owned by Archie MacGillis. He was considered one of the founding fathers of rum running working out of British Columbia waters. Leonard McCann Archives, LM2018.999.033, Vancouver Maritime Museum.

The shootout in Cadboro Bay was followed a week later by another Wild West encounter, this time with *Eva B*, which was disguised as a fishboat. It was owned by Seattle police officer turned rum running magnate Roy Olmstead. The capture of *Eva B* and its seven hundred cases of liquor by the *Winamac*, "a sturdy little craft, well suited for the work she was engaged," took place at Portland Island off Sidney on October 5, 1924. As it happened, Captain Abraham Reid Bittancourt (who was from a pioneering Salt Spring Island family), was chartering the 47-foot wood launch to the customs service as a patrol boat and operating out of Ganges Harbour on Salt Spring Island when he received word from fishermen of suspicious activity out at Portland Island. Following up on the tip, Captain Bittancourt and his son Lyndell ("Len"), *Winamac*'s engineer, donned their customs officer uniforms, grabbed their rifles and called out to local police constable J. N. Rogers and forestry officer A. Warburton to ask if they'd care to come along for a little excitement. They did.

Originally built as a US Revenue cutter in Marine City, Michigan in 1892, the 47-foot wood vessel *Winamac* went on the Canadian registry in 1909. Here, a man who is probably Captain Bittancourt is leaning out the wheelhouse window as the workboat steams out of Victoria's Inner Harbour. Drell and Morris Photographers, Ted Aussem collection.

After making the hour-long run over to Portland Island early Sunday morning, *Winamac* made a point to pass well abeam of the north side of the island, so as not to raise suspicion while still allowing the captain and crew to be able to look back and spot the three vessels anchored in a bay. Once around a headland and out of sight, Captain Bittancourt turned close in towards shore and cruised back into the bay before the boats' crews were even aware of their presence. One was a speedboat that was able to escape at some twenty-five to thirty miles an hour and get away to safety (*Winamac*, driven by a three-cylinder Corliss gas engine, was only good for nine knots at most), while "under a heavy fusillade of lead from the customs boat," the two other vessels, an "innocent looking" fishboat and *Eva B*, made no attempt to run and their capture was inevitable."

The *Daily Times* pointed out that "other customs officials state that it was one of the most adroitly made seizures ever made by

Captain A. R. Bittancourt, pictured here with his wife and two sons, was from a pioneering Portuguese family that settled Salt Spring Island. He was freighting among the Gulf Islands with his motor launch *Winamac* when the boat and crew were hired by the Canadian Customs and Excise Department. The "sturdy craft" was also under orders to the fisheries and immigration departments. Salt Spring Island Archives, no. 992101004.

local men, and all those accompanying him [Captain Bittancourt] are deserving of high praise." Especially since "the *Eva B* is one of the many high-powered launches operating in the waters of the Gulf of Georgia, running liquor from the Canadian to the American side, and as the territory is wide in which these boats are known to be present, and there being thousands of hidden little nooks where they might well avoid the eyes of a half hundred Government boats, it is a feather in the caps of the men who took her." (The "innocent looking" fishboat, which was having engine troubles, was set free since no liquor was found aboard.)

Later that day, *Winamac* returned to home base, Ganges Harbour, with her prize. There *Winamac* and the *Eva B* remained moored for the night with crew members retiring around midnight, leaving Constable Rogers to stand watch. The three American rum runners—*Eva B*'s master, Jack Rhodes; engineer Erickson; and "guest" Mr. Green (an ex-army officer who was actually the owner of the gas boat)—were taken up to the Ganges Harbour House for the night, where they were placed under guard. Around midnight, Constable Rogers noticed a strange launch slowly idling into the harbour, headed towards *Winamac*, apparently intent on coming alongside. He soon recognized it as the speedboat they'd surprised at Portland Island returning in an attempt to recover *Eva B*'s valuable cargo. Captain Bittancourt hailed the boat, giving it a warning, but still played it safe and kept his gun at the ready. But the occupants of the speedboat were just as quick to go for theirs and an exchange of bullets quickly broke out.

As Len Bittancourt recalled in 1965 to Imbert Orchard in a taped CBC Radio interview, "I thought I would get lots of sleep but then about midnight all the shooting that went on down here ... I don't know if they saw us but there was another vessel, the *Three Deuces* ... a high powered launch and come there to hijack her [unbeknownst to them *Eva B*'s cargo had been transferred to the hold of *Winamac*] ... but nobody got hit anyway. It was a little exciting here for a while; especially when you just nicely got into bed ..." Encountering a stiff defence, the interloper retreated from range to open up her engine and race off down the strait.

The *Three Deuces*, which had reportedly broken the Lake Washington speed record at one time, was operated by Prosper Graignic, one of Roy Olmstead's skippers, and propelled by a Liberty L-12 marine conversion—a four-hundred-horsepower V12 engine developed for the armada of fighters, bombers and observation planes built in the US during World War I. Graignic, who was described by Olmstead biographer Philip Metcalfe as "short and quiet," was rumoured to have made almost four hundred trips for Olmstead by the spring of 1924.

Len Bittancourt explained in his interview how rum running was supposed to work—that is, if one was a Canadian citizen. Canadian boats were allowed to load a cargo and clear for a given point. "They're not breaking the law ... but [they did] break the law ... when transferring it to an American vessel without supervision of a customs officer ... As far as we were concerned they weren't renegades and another thing, there was so many of them you couldn't keep up to them! ... But we did get the odd American."

The *Daily Times* of October 7 reported that the morning after the Sunday night shootout in Ganges Harbour, *Winamac* brought the *Eva B*—"a fine, seaworthy craft, with a good cargo capacity"—into Victoria's Inner Harbour. Rhodes, Erickson and Green were promptly brought up before a customs department board of inquiry and let out the following day on a $250 cash bond. Since they were unable to explain satisfactorily why they were in Canadian waters, they were turned over to the immigration department for deportation. But then there was the question of the mixed cargo of some seven hundred cases worth about four thousand dollars. (Around fifty-six thousand dollars in today's money.)

There was some discussion following the seizure as to what was to be done with the captured alcohol. It was suggested that in the future when Provincial Police assist, all liquor be promptly destroyed on the spot, thus eliminating any fear of future attacks from hijackers. But in this case, which resulted in considerable financial benefit to the local customs department, the Provincial Police were rather annoyed since they didn't see any of it, even though they took part and would have been willing to pay the regular customs duties.

Still, while it was the shenanigans occurring throughout the province's southern waters just outside the large urban centres of Vancouver and Victoria that were grabbing the population's attention, the waters off the west coast of Vancouver Island and along the entire BC coast right up to Alaska were to serve as major liquor entrepôts right up until Prohibition was brought to an end in 1933.

On October 22, 1924, a story titled "Seizure is a Mystery" in the *Vancouver Sun* noted that the position of the "liquor carrying boat *Impala*" upon her capture was somewhat baffling. (The *Impala* was a 47-foot, 8-inch wooden boat built in Vancouver in 1912.) It appears that Provincial Police were at a loss to understand what a liquor runner was doing out in Juan de Fuca Strait as far north as Bamfield since "the fleet has centred its operations in the Gulf of Georgia with bases on the numerous lonely islands off Victoria." The discovery of a liquor boat off the west coast had them completely puzzled, and in particular, "where the *Impala* got her 483 cases of liquor and where she intended to take it are both questions which the police have been unable to answer so far." The reporter went on to note that "the case, however, is being investigated carefully in an effort to uncover what is thought to be some new plan to outwit the Canadian Customs and the US Prohibition law. It was thought that perhaps the liquor smuggling fleet alarmed at the sudden new effort of the Canadian authorities to curb their operations may have decided to shift its base to a point far removed as practicable from the usual liquor running waters ... the fact that the Fisheries Patrol Boat *Givenchy* was used to seize the *Impala* shows that all branches of the Federal Government service are cooperating in the attempt to curb the liquor fleet and its utter disregard for Canadian as well as American law." It concluded with the note that while the Canadian authorities hadn't decided as yet what was to be done with the *Impala*, "her captain and crew have told the Customs Authorities here that they are not guilty of violating any Canadian law."

Still, it's rather perplexing as to why Canadian customs was surprised by this development nearly four years into Prohibition, since there was already a flourishing liquor trade going on off the west coast of Vancouver Island. BC resident Richard Thompson recalled that when he was growing up, his family always wondered how it was that their grandfather never seemed to have to work. They only began piecing together the story of how their grandfather was able to retire early in life when they realized that he'd run a store at Spring Cove near Ucluelet from 1912 up until 1924. It seems that old Grandad had done rather well serving the rum running fleet in the early 1920s. Apparently, many a time boats showed up at the store's float in the middle of the night and cleaned out all its stock.

Some accounts also have it that the proprietor of the village of Clo-oose's store, just down the coast from Spring Cove, also provided a friendly service to the rum running fleet from her "thieves' market" and was so successful that she was able to retire to a Vancouver hotel by the late 1920s. These particular "thieves" would have been the smaller vessels that were making runs to the mother ships out in international waters in the open Pacific off Cape Beale and Cape Flattery. Since the mother ships were required to sit out on the spot for weeks at a time, they most likely had fish packers or towboats that were delivering or picking up liquor stop in at Spring Cove and Clo-oose to replenish grub, oil and supplies. The west coast of Vancouver Island and the open waters of Juan de Fuca Strait were to remain a popular location for passing over liquor to American vessels right up until the end of Prohibition in 1933. Still, as Johnny Schnarr noted, there was a downside. By the early 1930s, there were so many American boats running out Juan de Fuca Strait to load, it drove the price down to five dollars a case. Up until that time, the Canadian boats had been getting eleven dollars a case.

Following the seizure of *Impala*, Victoria's *Daily Times* reported on October 21, 1924, about "captured boats increasing in harbour waters"; the fleet of seized ships included *Impala*, *Eva B* and *M-322* (the original account of her capture in July erroneously identified her as *M-332*). After noting that the *Impala's* cargo was principally whisky and gin, those who inspected the freight she carried still had to decide whether she was guilty or not. Since Prohibition came to an end in British Columbia in 1920, the movement of liquor within Canadian waters was entirely legal. That is, as long as customs regulations were followed and, if the liquor was to be exported, the Canadian federal government paid its due. Otherwise Canadian officials tended to become somewhat irritated.

The customs department continued to maintain a watchful eye on rum running marine traffic. On October 14, 1925, they seized the packer *Kiltuish* as she arrived at the William Head quarantine station just outside Esquimalt harbour, apparently after having unloaded her liquor cargo at sea. The 87-foot, 7-inch fish packer was registered to Arctic Fur Traders Exchange Limited of Vancouver, and owned by Archie MacGillis and Frederick Rae Anderson, barrister. When customs officers inspected the vessel, they were rather annoyed to discover she had sailed for a foreign port with only coastwise clearance papers. As *The Daily Colonist* pointed out, regardless of the fact that steamers that were plying between Victoria or Vancouver and California were generally spoken of as being in the coastwise service, this didn't mean that vessels could clear for California ports with coastwise papers. According to customs regulations, any port outside of Canada, whether it was down along the American west coast or in another country, was considered foreign.

After paying a four-hundred-dollar fine for the infraction, *Kiltuish* was released to continue on its way to Vancouver, and in 1927 joined the fleet of Consolidated Exporters, the big-time liquor export consortium working out of downtown Vancouver.

The *Kiltuish* was just one in the fleet of small boats and operators willing to reap handsome rewards through rum running in southern BC waters during the US Prohibition years. On January 24, 1925, it was reported on *The Daily Colonist*'s Marine and Transportation page that the two-masted auxiliary schooner *Lira de Agua* was in berth at the Ogden Point docks, loading a large consignment of liquor bound for Corinto, Nicaragua. Shortly after returning from her voyage to southern waters, the vessel was seized off the west coast of Vancouver Island by Canadian customs officers. She was alleged to have transferred one thousand cases of liquor to the *Ououkinish*, another two-masted auxiliary schooner, while still in Canadian waters. Because the transfer was made in Canadian waters, following her alleged delivery to a foreign country, it seemed likely that she didn't have the required coastwise clearance paperwork aboard for doing so.

In October, later that year, the Victoria daily reported that *Lira de Agua* was to be confiscated by the Government and eventually sold, provided that the case was not appealed by the owners. *Lira de Agua*—which at one time had been used in the making of Hollywood films—was originally purchased in 1922 by R. M. Morgan and Company of Vancouver, to all intents and purposes for "coast freighting" as Mr. Morgan informed a *Vancouver Sun* reporter at the time. He also pointed out that he had no intention of changing her Nicaraguan registry since she was to ply up and down the coast between Vancouver and Fasceta Bay, Nicaragua. Shortly thereafter Morgan sold the schooner to Northern Freighters, which was owned by the well-known local brewery and distillery interests, the Reifel family. As for the 56-foot, 4-inch *Ououkinish*, a former halibut schooner, she was registered to the Atlantic & Pacific Navigation Company, another shipping subsidiary owned by the Reifel interests. The *Colonist* noted that the vessel was seized over on the mainland for her role in the transgression.

Also, that fall of 1925, the two-masted auxiliary schooner *Chakawana*, with a crew of five aboard, was seized off the west coast of Vancouver Island, this time by customs officers aboard the Royal Canadian Navy training ship and occasional fishery patrol vessel, HMCS *Armentieres*. The 62-foot *Chakawana*, which was owned by Archie MacGillis's Coal Harbour Wharf and Trading Company, was brought in for having liquor on board that had not been loaded under customs jurisdiction, but was released on

In April 1923, the *Vancouver Sun* reported that R. M. Morgan was to use *Lira de Agua*, a schooner that had been used in the making of Hollywood films, for coast freighting. Leonard McCann Archives, LM2018.999.034, Vancouver Maritime Museum.

security. The cargo consisted of nine hundred cases of various liqueurs and whisky.

In his book *Slow Boat on Rum Row*, Fraser Miles tells of his first voyage aboard the fish-packer-cum-rum-runner the 61-foot, 7-inch *Ruth B* in December 1931. They made transfers at contact points off Cape Beale but he noted that they ventured as far north as the Queen Charlotte Islands (today's Haida Gwaii) to sit out off Alaska waters.

Here they would wait for a radio call from Vancouver ordering them to release cargo to American boats. While nowhere on the scale of the smuggling that was being carried out down south, there was quite a flourishing trade working out of Prince Rupert and nearby waters. Olier Besner, one of Prince Rupert's most colourful characters, jumped right in when the Volstead Act was enacted and set up a "fox farm" out on the Kinahan Islands just outside the entrance to Prince Rupert. For all intents and purposes, the "farm" served as a depot for exporting liquor into Alaska. In his biography, *Charlie's Tugboat Tales* (written by Prince Rupert resident Bruce Wishart), Captain Charlie Currie, who was working on a pile driver nearby at the time, described how Besner ran his export terminal: "Well, they had a place you could back right in there in the dock, and a little winch and a derrick ... It was quite often one of those big American seiners would come in there, back in and just load up right full of booze. It was never in boxes, all of it in sacks that would sink if they thought they were gonna get pinched ... they had signs around the beach at Kinahan: Fox Farm. That was just a blind you see ..." Unfortunately, Besner was being hijacked regularly and even tried sinking sacks in the ocean to evade the hijackers, but still, they posed enough of a problem that he decided to quit the business. In the end though, it seemed he came out of it all right financially. In 1928, Besner built a beautiful building on Prince Rupert's main street in the Spanish Colonial Revival style, the Besner Block, which still stands to this day. He also owned the Besner Apartments and was a charter member of the city's board of trade.

Overall, there were countless individuals in the rum running trade who became very adept at eluding both American and Canadian law enforcement agents, along with the criminal element, in order to earn a decent and often very lucrative living. As Hugh Garling said, "For every boat seized, the smugglers reaped $1,000 from successful sales, and for every man arrested hundreds more ignored the law." While the local cross-border trade in illegal liquor was a dangerous business fraught with risk, there was also another method of freighting into the American market that was more benign, a daylight operation that was generally quite open and legal. Large mother ships loaded liquor from bonded warehouses and then made deep-water voyages to sit off the coast of California in international waters or in later years to anchor out just south of Ensenada, Mexico. Here, they supplied smaller freighters that delivered orders to buyers' boats, which then assumed all the risk by having to transport the illegal cargo into out-of-the-way beaches along the American coast.

MOTHER SHIPS AND THE DISASTER YEARS (1923-24)

*Just outside the territorial limit they all lie, riding at anchor, great, rust-streaked, iron
steamships, once-palatial ocean-going yachts of multi-millionaires, stout, heavily-built,
powerfully-engined, deep-water towboats, lofty-sparred, noble, old ships and barks with
wide crossed yards, salt-encrusted tramps, and all filled to the hatches, laden to waterline
or plimsoll, with their cargos of liquor from every quarter of the globe—rum from the West
Indies, gin from Holland, whisky from Scotland, wines and brandy from Spain, Portugal
and France, beer from Panama, Chile and Germany ...*

—A. HYATT VERRILL, *SMUGGLERS AND SMUGGLING*

A substantial amount of the rum running business, and perhaps the most rewarding, took place far from coastal BC
waters. The deep-sea liquor trade was the stuff of legend,
with mother ships—deep-sea steamers along with retired lumber and sealing schooners—that travelled south from ports such
as Vancouver and Victoria to sit well out in international waters
off the Oregon and California coastline or, in the later years of
Prohibition, to anchor not too far south from the small Mexican port
town of Ensenada. Companies such as Consolidated Exporters and
Archie MacGillis's Canadian-Mexican Shipping and Roy Olmstead's
Western Freighters had mother ships sitting at anchor or drifting off
the US west coast. They acted as large sea-going liquor emporiums

that supplied the smaller vessels that ran their orders into various points along the coast.

Hugh Garling went into more detail on how it all worked: "the [motor ship] *Lillehorn* would be the mother ship until it was time for her to return home. The [five-masted auxiliary schooner] *Malahat* would relieve her and the *Lillehorn* would transfer the balance of her cargo to the *Malahat* which would arrive with a cargo of liquor. She would stay there for four months to a year or more. These were the larger ships that could stay at sea for long periods of time and would supply the next in the hierarchy, distributor boats [usually large fishboats or packers], with liquor, provisions, diesel oil and even food supplies. The primary function of the distributor boats was to transport their load of liquor up to various agreed to positions off the coast where buyers' boats would come out to pick up their pre-ordered liquor. The buyers' boats came last in the hierarchy and were small, fast boats, usually American, which would make contact, identify themselves and take their prepaid load, landing it on the beach, sometimes running it right in a harbour, through the Coast Guard blockade."

But still, often Canadian "shore boats" took on a big risk and loaded from a mother ship to run a load of liquor across the line directly into American waters. In researching his memoir, Fraser Miles gained access to a 1977 taped interview of crewman Clarence Greenan, who was with *Malahat* in her early years operating as a mother ship off the California coast. Greenan, who sailed on *Malahat* for five years as donkeyman and shore boat runner, was only twenty-one years old when he finally signed off the ship in 1924. He made two hundred dollars a month plus a trip bonus of two months' pay while crewing on the mother ship. After he quit the racket some three or four years later, he claimed he took home a sizeable stake of twenty thousand dollars. Most of this, he pointed out, was for driving *Malahat*'s fast shore boat inshore all the way up the Sacramento River to a private dock at the state capital once or twice a week, for which he was paid $450 a trip. Here, he would land at a private dock and, after unloading, would gas up to replace the five hundred gallons burned through making the run. Of course, these deliveries were all done at night. He would cast off from *Malahat* just before dark and be back alongside the big schooner just after daylight.

In 1984, George Winterburn, a crewman aboard the five-masted auxiliary schooner turned mother ship *Malahat*, explained how it all worked to BC writer Susan Mayse, whose tape of the interview sits in the BC Archives. "Several launches came out to us and upon presenting their credentials, consisting of one half of a one dollar bill that had to be matched with the other half that the supercargo kept, the order which was written on his half bill was filled and away he went to shore with his launch full of liquor." Land agents ashore collected cash payments in advance from American bootleggers and an order number was written on both sides of the dollar bill. Then, once enough orders and money had been collected to account for half the cargo, a mother ship was able to leave Vancouver. The whole operation was a straightforward shipping operation once the liquor was loaded in Burrard Inlet from bonded warehouses aboard steamers and sailing ships.

Mother ship captain Charles Hudson noted that while the rum running industry turned into one of the most fabulous money-making operations in Canada at the time, it still involved a lot of tedious shipboard work. Besides the routine duties such as washing down decks every morning, sail and rigging maintenance and repair, scraping marine growth off ship's bottoms, tarring decks and so on, they also had to deal with their substantial payloads. This was particularly labour intensive for the crews of the mother ships, since they devoted so much time to "sacking." In order to safely and quickly manhandle cargo over to small boats running alongside, the twenty-four-bottle crates' tops were broken open with an axe and their contents sewn up in gunny sacks. The crates' packing was used

Working aboard a mother ship required a lot of hard labour. Here the crew is transferring a load of liquor cases aboard the mother ship *Malahat* after the schooner *Marechal Foch* arrived on Rum Row, Ensenada, from Tahiti in the early 1930s. Hugh Garling collection.

to protect the bottles while each sack was labelled with a special code to identify the various brands it contained. Also, they were all sewn up with two ears, or lugs, on top to make the sacks easier for handling and for throwing down into the boats. There was also another benefit to sacking the bottles, besides easier handling and loading. If there was trouble in the air—for example, if a US Coast Guard ship was spotted nearby—the sacks could be dumped and laid on the bottom of the ocean on long lines to escape detection.

The remains of the broken boxes were thrown overboard or saved to fire the boilers of desalination condensers to produce fresh water. Writer Philip Metcalfe said that, "From Cape Flattery, Washington, to La Jolla, California, the empty cases washed ashore, their sides bearing the names of distilleries in Glasgow and London."

Local newspapers, like Victoria's *Daily Colonist*, were diligent in keeping the public up to date on the movements of the fleet. On September 23, 1924, it was reported on the Marine and Transportation page that "*Stadacona* Due Back Wednesday to Load Liquor for South—*Malahat* Is Expected Next Week and Will Load Here and at Vancouver—*Principio* Is Taking Out 25,000 Cases of Liquor."

It also noted that "these [three] vessels are operated by a consolidated company, in which considerable Victoria capital has been invested." (James "Jimmy" Hunter, president of Consolidated Exporters Corporation, managed the Victoria branch of the operation from the Pither & Leiser building in the Inner Harbour.) All three mother ships were to take on the Thanksgiving supply of liquor and return again in November. The Christmas supply was already on its way out to British Columbia from Europe "in a steady stream of consignments aboard Canadian Government Merchant Marine freighters, Harrison-Direct liners and Royal Mail Steam Packet ships. *Canadian Freighter*, which left the Old Country on September 8, has one of the largest shipments of liquor ever brought from the United Kingdom."

The 247-foot, 6-inch *Malahat* was perhaps the most successful of the mother ships and years later was to be recognized as the queen of the rum runners. On June 30, 1923, *Malahat* was loaded with booze and ready to depart on her first voyage in the lucrative trade. A wire report from San Francisco appeared on the Marine and Transportation page of *The Daily Colonist* under the headline "Rum Ship Anchors at San Francisco" on July 19. It reported that "a big vessel, believed to be the mothership *Malahat*, is at anchor beyond the three-mile limit off San Francisco, and is supposed to have supplied liquor to a swarm of smaller boats, according to an announcement today from the headquarters of the district federal prohibition enforcement officer here. The vessel left Vancouver, BC, for Mexico ... While the supposed liquor was being transferred, the small boats formed virtually a continuous line between the ship and the shore, and operated during the daytime as well as at night, the prohibition office was informed."

The retired lumber schooner was well suited for the trade since she was capable of taking on sixty thousand cases of liquor. When she put to sea, her hatches were plugged full and decks packed with case upon case of the choicest brands from Canadian and European breweries and distilleries. As Hugh Garling, who crewed on *Malahat*, wrote in one of a series of articles on rum running that ran in *Harbour & Shipping* magazine between 1989 and 1991, "Canadian-produced bourbon whiskey and Canadian rye, were the most in demand, but large quantities of famous proprietary brands of scotch, gin, rum, brandy, and well known vintage champagnes, sparkling burgundies, still wines and renowned liqueurs, were carried. Mother ships might carry up to 150 or 175 different items."

While *Malahat* made it through the Prohibition years unscathed, the West Coast's deep-sea liquor trade wasn't exactly trouble-free or without serious incident, including losses due to marine mishap as well as seizures on the high seas. While Fraser Miles claimed in his autobiography that the years 1922 and 1923 were "the fabulous years of rum running" since it was still a relatively simple, presumably profitable and trouble-free undertaking, he did go on to say that the next two years, 1924 and 1925, were "the years of disaster." Miles attributes the losses during this period to "error, carelessness or overconfidence on the part of the rum runner ship captains involved, of which the US Coast Guard was happy to take advantage."

Miles claimed that the two-masted auxiliary schooner *City of San Diego* became the very first vessel to make a rum running voyage when she left a BC port with her holds filled with liquor on April 15, 1922. But what Miles may have been suggesting was that the vessel was likely the very first mother ship to head out and sit in international waters just off the Washington, Oregon or California coastlines. As it happens, *City of San Diego*, which was built as a sealing vessel and launched from the J. Turner shipyard in San Francisco in 1881, was 67 feet, 6 inches in length and registered to the Reifel interest's Northern Freighters.

But the first mother ship to make big headline news up and down the coast was actually a vessel that had loaded and sailed from a European port, not a British Columbian one, though her ultimate destination was Victoria. After departing Leith, Scotland, with twenty-five thousand cases of Scotch whisky early in 1923, the British steamer *Ardenza* (210 feet in length, launched in 1920 from the yard of Hawthorns & Company in Leith) passed through the Panama Canal and steamed up the California coast where she sat for nearly eight months before arriving in Victoria after a ten-month voyage.

In its June 1, 1924, edition, *The Daily Colonist* devoted five columns to the inside story on "mystery ship" *Ardenza*. After detailing the flight of Mr. C. T. Steven, the ship's owner, from Scotland to Argentina to escape his creditors, the reporter went on to tell a tale "which set the whole Pacific Coast agog with interest. Shades of Captain Kidd, of Bloody Morgan, and all that company of rollicking swashbucklers, must surely have hovered delightedly above the rum-running Glasgow freighter, during her fantastic voyage."

The article reported that after arriving in Victoria "without a scrap of paper as evidence of her registry, her destination, or the nature or quantity of her cargo, her career was shrouded in mystery." That is until Able Bodied Seaman Murchison, "disgruntled with his lot in life and harbouring a grudge against his captain and shipmates, under the influence of Bass ale in a waterfront grogshop, lifted the veil of secrecy." Murchison told of how the ship's papers were thrown overboard, and of the drunken carousals and fierce fights that marked the voyage, as well as mutiny and even attempts to wreck the steamer. He also went into some detail explaining how the whole rum running business operated off the California coast and how the owners of *Ardenza* hoped to benefit from the lucrative trade.

"We left in May last year with about two dozen cases of gramophones and case goods mixed in with 27,000 cases of whiskey, to give us the right to clearance with a general cargo for Vancouver." He said it was rather a prosaic voyage at first, but things went downhill once they were off the California coast. When they arrived off Los Angeles, no small boats came out to pick up orders since word had been received that the "prohis" (Prohibition officers) there couldn't be fixed satisfactorily, so they steamed up the coast to sit off Half Moon Bay. Here, conditions appeared better since San Francisco's prohis were easily persuaded for $2.50 a case "to put the telescope to the blind eye" and look the other way. But troubles arose when "bibulous San Franciscans" refused to drink *Ardenza*'s product. While the whisky bottles were indeed affixed with quality labels reading "Johnnie Walker" and "Black & White," discerning imbibers were quick to catch on that while it was undoubtedly whisky, it was of "a quality never to be found on the table of a Highland connoisseur." And where did they obtain these labels? To start with, Murchison said that there was no shortage of printing presses in San Francisco where forged labels could be run off. He added that the crew spent a long time aboard the steamer opening cases and changing labels.

At first, the volume of business was good, with the average price of a case, which probably only cost five dollars in bond to ship from Scotland, going over the side for ninety dollars. "All manners of small craft ventured out at night to *Ardenza* ... manned by men armed to the teeth, the biggest and toughest rogues in San Francisco ... hard-boiled gunmen from the Barbary Coast. For hijackers infested the waters and the game was dangerous. More than one battle was fought within full view of the Glasgow ship." Murchison said that rum runners "outside the ring" were having a rough time of it trying to sell into the San Francisco market because of the competition from Canadian mother ships.

After three months of struggling to make a go of it, sales began to fall off dramatically since their competitors, *Prince Albert* and *Coal Harbour*, were cutting prices. The cost of a case went down from ninety dollars to eighty and then seventy dollars, and still kept dropping. Also, once the news was out in the bars and taverns of the city that *Ardenza*'s cargo was cheap stuff and her British Columbia rivals were peddling the quality brands, buyers' boats stopped coming alongside. As a result, conditions on the ship worsened. Soon the crew were subsisting on "salt horse" while fresh water supplies ran dangerously low. Then the North Pacific gales rolled in and mutiny was in the air while night after night the crew continued to filch cases of whisky from the hold and all (with the exception of Captain Whittle and one of the engineers, a teetotaler) would lie around in a drunken stupor. Broken heads and lacerated features quickly became common occurrences.

In February of 1924, several weeks after abandoning the failed venture and arriving in Victoria, *Ardenza* was seized for debt and sold to new owners. The steamer finally obtained provisional papers from Ottawa and was to return to Scotland, but not until after she unloaded the 7,400 cases of unsold whisky cargo in the capital city. *The Daily Colonist* reported on the twenty-sixth of that month, "It will go into bond to await disposition by its owner." After proceeding to Union Bay to fill coal bunkers, the ship returned to Victoria to load four to five hundred thousand feet of lumber from Victoria's Canadian Puget Sound mill as cargo for their return voyage to the United Kingdom. Later that summer, with *Ardenza* long departed, the *Colonist* added a footnote to the tale. It had received word that the disastrous outcome of the steamer's failed venture was probably the leading cause that led Sir John Stewart, the Scottish whisky baron who sent the "mystery ship" out on her fateful voyage to the Pacific Coast, to blow his brains out at Fingask Castle.

Overall, though, up until the fall of 1924, the deep-water liquor export shipping business working out of British Columbia ports appeared to be rolling along quite well and without undue problems. There had been a capture in May of that year by the Mexican fisheries patrol vessel *Tecate* off the Coronado Islands, but the seized *West Coast* was only a small two-masted schooner of 55 feet in length. Since the vessel had no Mexican clearance papers, boat and cargo were confiscated. Registered to the Red Star Navigation Company of Vancouver, the crew consisted of only the skipper and his wife.

The seemingly well-oiled liquor distribution system began to unravel when the steamer *Quadra* was seized off San Francisco on October 12, 1924. Second Engineer George Winterburn described what transpired in Ruth Greene's *Personality Ships of British Columbia*. "We loaded up in Vancouver with 22,000 cases of choice liquors, wines, rums and a large quantity of beer which was all consigned to Ensenada in Mexico. Papers were arranged to show that we had been there, discharged our cargo and left again with a clear bill of health. This we had done before we even left Vancouver."

As it happened, it was the *Quadra*'s first voyage in the liquor-carrying trade. She was working out of Victoria and Vancouver after undergoing reconditioning at the Victoria Machinery Depot. Originally purpose-built as a lighthouse and buoy tender, the *Quadra*, 174 feet, 6 inches in length, was launched in 1891 in Paisley, Scotland, as a steel screw steamer for Canada's Department of Marine and Fisheries. A 120-horsepower quadruple-expansion four-cylinder engine drove a single screw. She had also doubled as a fisheries patrol vessel and, with these government duties (the crew wore uniforms and were well disciplined, much like the crew of a naval vessel), served as symbol of law and order up and down the coast. In one of his *Harbour & Shipping* articles, Hugh Garling pointed out that she was easily recognizable and a particularly distinctive vessel. "Her clipper bow, the bowsprit, the overhanging counter of her elliptical stern, her raking masts and funnel, gave her a most yacht-like appearance." After being retired from government service, the *Quadra* had been purchased by Britannia Mining and Smelting Company to transport ore from Howe Sound to a Tacoma smelter. In September 1924, Archie MacGillis's Canadian-Mexican Shipping Company purchased *Quadra* from Britannia Mining and chartered the steamer to Consolidated Exporters. *Quadra* wasn't even out of the Strait of Juan de Fuca that October when they stopped for their first night of business.

From Juan de Fuca, *Quadra* sailed south to sit out off the mouth of the Columbia River, where they carried on a brisk business with the suppliers to the Portland market. They then left for San Francisco and took a position in supposedly international waters just outside the Farallon Islands and fifty miles offshore from the Golden Gate. As writer Ed Starkins describes, this location served as "a virtual open sea marketplace for smuggled liquor" throughout the early 1920s until it became too "hot" with US Coast Guard cutters. Unfortunately, *Quadra* was forced to sit out a bad storm for a week, but still there was no rest for the crew since they were kept busy "sacking" liquor.

When the mother ship *Malahat* appeared on the scene to deliver more cargo to the *Quadra*, the *Malahat*'s captain tried to bring her alongside. However, he misjudged the distance and accidently rammed her, taking the steamer's bowsprit right off and punching a sizeable hole in her side. Bob Gisborne, who was crewing on *Quadra*, told Hugh Garling that his cabin was right up forward and after the collision he could almost walk right out into the sea. He said that the *Quadra* would have gone down in a Pacific blow. Then, while both crews were engaged repairing the breeched hull by stuffing it with mattresses to keep the sea out, they failed to maintain

The *Quadra* just after her capture. Note the missing bowsprit, which was broken off in a collision with another mother ship, the *Malahat*, out on Rum Row off the Farallon Islands. US National Park Service, B05.19408.

their usual close watch on the waters around them. They soon realized they'd made a grievous mistake when they found themselves looking down the gun barrel of the US Coast Guard cutter *Shawnee*'s twelve-pounder.

As it happened, the *Shawnee*, which was stationed at San Francisco and responsible for patrolling from Cape Blanco, Oregon, down to the Mexican border, was cruising offshore keeping a sharp lookout for the 190-foot freighter *Principio* when it received word that *Quadra* might be about too. The *Shawnee* was a Coast Guard auxiliary steam tug, 158 feet in length with a cruising speed of ten and top speed of twelve knots. With Lieutenant Commander Charles F. "Shorty" Howell in command, she was six miles northwest of the

The us Coast Guard cutter *Shawnee* lies at anchor off Sausalito on August 24, 1936. US National Park Service, B06.38479.

Farallones when they spotted a vessel off on the horizon. Steaming around to seaward to cut off a possible escape to open waters, they soon identified the vessel as *Quadra*. *Shawnee* quickly went in pursuit and, once the rum runner was overhauled, gave four blasts of her whistle signalling for the mother ship to stop. A small fishboat, *c-55*, which just happened to be caught alongside after loading some fifty sacks of liquor, attempted to escape but after *Shawnee* sent a shot across her bows, gave up and surrendered. It is likely that the *Malahat* slipped away during the action.

In Ruth Greene's book, *Personality Ships of British Columbia*, George F. Winterburn, another *Quadra* crewman, said he would never forget the incident. "He fired a warning shot at a launch which tried to slip away. That proved conclusively to us that he meant business and would not take a few cases of liquor to let us go." Third Mate and Bosun Bob Gisborne told Hugh Garling, "She fired one shot, he'd [the *Quadra*'s captain, George Ford] never been in the war before ... I'd been shot at, but he wanted to stop but I said 'he'll never hit us.' Next shot lifted the stern right up, so the skipper decided to stop then ... They came alongside and wanted us to take the [tow] line." Winterburn said that he hid seven cases of whisky down the double-hull tank in the engine room, and then flooded the tank in the event the cargo was seized and the ship returned. "We would at least have some fortification against melancholia or sea-sickness, but alas we were taken ashore next morning and I never laid eyes on her since." Since Captain Ford refused to give his name or show his manifest and wouldn't get steam up for San Francisco Bay, a prize crew was put aboard. *Quadra* was then taken in tow by the cutter along with the "fireboats" *C-55* and *D-905*, captured while standing off waiting to load from the *Quadra*. Short for "fire water boats," fireboats were the small, fast boats that would come out from US beaches and ports to pick up deliveries from Canadian vessels. Captain Ford, an ex-Navy officer, protested their arrest on the grounds that they were more than an hour's steaming time from the US coast, therefore well out in international waters. It was to no avail.

Lieutenant Commander Howell credited his good fortune in making the sensational seizure to the rum runner's failure to receive a perfectly innocent wireless message that somebody's grandmother was about to die. Explaining, he said, "These wireless messages are sent from amateur shore stations. The rum-runners never send one in return. They merely receive. Innocent messages like 'Grandmother died today,' or 'I am going to the ball game,' fill the air every time the *Shawnee* gets ready to put out. That is why we are rarely able to find anything. They know we are going out before we start and they just steam out of sight until the coast is clear."

When *The Daily Colonist* broke the story of the *Quadra* seizure on October 12, 1924, it said, "It is claimed on good authority if Captain Ford was able to prove that his cargo was destined for Mexico and none had been touched, then *Quadra* would be released. It was pointed out, however, that under the new British-US treaty signed earlier in the year no ship carrying liquor was allowed within the twelve-mile limit unless it was under the British flag and the cargo sealed." Lieutenant Commander Howell's report claimed that *Quadra* was well within the County of San Francisco, being only seven miles northwest of the Farallones. Captain Ford argued otherwise. He pointed out that he and two officers had just taken a sextant reading that set his position seventeen miles west of the island group, and there was no way *Quadra* could have steamed into a California port within an hour at the time they were seized. In the initial judgment given on January 5, 1925, United States District Judge John S. Partridge repeated the relevant provisions of the treaty. Probably the most pertinent provision was the one that stated that the one-hour steaming distance, used to determine whether an offence had

In writing to the Commandant of the Coast Guard, the us attorney noted that the clean-cut, fine appearance of Captain Charles F. Howell and his men as they appeared in court, and the straightforward manner in which they testified, was a great credit to both the Coast Guard and the government. Archival photo published in Malcolm F. Willoughby, *Rum War at Sea*, United States Coast Guard, 1964.

been committed, was that of the vessel conveying the liquor into the United States: the fast American launches running into shore, rather than the slower mother ships that were actually boarded and searched.

While Bay Area residents were eager to get down to the Embarcadero and get a first-hand look at some Canadian rum pirates being brought to justice, Captain Ford let his crew know that while they were probably headed for jail, they were not to worry as arrangements were being made to get them out on bail as soon as possible. In order to achieve this goal, he ordered them all to shave, clean themselves up and put on their best clothes, since it was all going to be part of the show. So contrary to what the crowd was expecting, they were surprised to see some twenty to thirty pleasant-looking, clean-cut young men coming down the gangplank.

As Captain Ford promised, the very next day officers and crew were released on bail on bonds totalling $175,000 while their cargo of what was reported to be some twelve thousand cases of whisky and champagne was placed under federal guard. Following his crew putting on a fine display of their best behaviour for the American public, Captain Ford informed local reporters that two of *Shawnee*'s crew got quite drunk while on board *Quadra*. One even became so inebriated he had to be put in irons by the cutter's officers.

After *Quadra*'s crew was released on bail, Consolidated Exporters paid them full salary and shore maintenance (most likely room and board) in order to keep them on hand for when they were called to testify and give evidence in court. As a result, they got to enjoy time ashore for eight months while collecting a wage estimated to be double that of the standard mercantile pay of the day. Rumour had it that nearly all of them appeared to be broke by the time the next payday rolled around. Ruth Greene quoted Charlie Coppins, a mate on *Quadra*, about his experiences once ashore. "We lived like lords in San Francisco. We had the best food, lived in hotels, and had to report to the authorities during the day." Still, Charlie had some explaining to do since he had told his wife that he was going off fishing. Bob Gisborne did all right himself while waiting it out in San Francisco. He quickly settled into a regular routine: in the morning he'd go play pool, in the afternoon go for a swim in the public pool and then in the evening maybe take in a dance. He said there were lots of parties but since he didn't drink, he didn't bother going to any. On the other hand, some of the crew had their wives come down to 'Frisco for a holiday. Gisborne recalled visiting

Captain Ford when he was initially locked up in a cell. "Did you know they had sheets on the bed? We never had sheets on the boat, and bacon and eggs for breakfast ... he had a table with a nice cloth on it ... and bananas and oranges on the table. And this is supposed to be jail."

Hugh Garling learned that Donald McDonald Donald, who was also crewing on *Quadra* at the time, put his paid idleness ashore to good use. He purchased course books, hired a master mariner as a tutor and proceeded to study for his certificate. Before the *Quadra* trial ended, he'd passed his examinations to qualify as master, coastwise. While he was entirely grateful to running for where it got him, he still took a dim view of the trade and its effect on many who served on the rum ships. "Rum runnin' spoils a man's morale and might tend to make him dissatisfied wi' ordinary wages when he has to go back to work. So I thought I'd get out before I got wedded to it. 'Tis an enticin' lazy pastime."

On January 5, 1925, Judge Partridge handed down his decision that the seizure of *Quadra* "with its supposed cargo of illicit liquor, valued at $500,000, was regular in every way, and not in contravention of any treaty" according to Victoria's *Daily Colonist*. Objections to the admitting of some of the evidence were rejected and the trial proceeded. Along with the captain, officers and crew of the *Quadra*, Vincent Quartararo and Charles H. Belanger were also charged with conspiracy to violate the National Prohibition Act and the Tariff Act of 1922. (The Fordney–McCumber Tariff Act was a law that raised American tariffs on many imported goods to protect factories and farms.) Quartararo, who was based on shore in San Francisco, was claimed to be the most active agent of the conspiracy working for Consolidated Exporters while Belanger served as its director in San Francisco.

It was charged that while sitting off the Farallones, the captains of the mother ships *Malahat*, *Coal Harbour* and *Quadra* were in constant contact with Quartararo and Belanger and, to some extent, acting under their orders. It was discovered that Belanger, as the Bay Area director for Consolidated Exporters, arranged for and had sent out to *Malahat* the burlap sacking used to pass over the liquor to *Quadra*. US government evidence found that Belanger was also the one giving the orders to transfer liquor from one vessel to another and even to transport designated liquor from ship to shore. He also made the arrangements for *Quadra* to be supplied with fuel oil from the Bay Area. When Chief Justice Taft finally delivered the opinion of the court, on April 11, 1927, some two years later, it was pointed out that "none of the three seagoing vessels proceeded to their destinations officially described in their ship's papers, but cruised up and down between the Farallones and the Golden Gate, where the exchanges of liquor and sacks were made, and where the needed [fuel] oil was delivered, and from which the liquor was carried by small boats to a landing place called Oakland Creek in San Francisco."

As the trial proceeded, it soon proved a great source of fascination and amusement to both the Canadian and American public. On 15 March 1925, *The Daily Colonist* reported that Federal Judge Partridge's court "took on an appearance of an old time saloon when customs officers wheeled evidence into court tank after tank of beer, barrel after barrel of whiskey, case after case of wines, cordials, champagnes and rums ... All that was missing was the brass bar ... Jurors gasped, court attaches mumbled 'Aaahs' and 'Ooohs' and the court audience generally smiled ... It was a pleasing sight for some and a scene that they had not witnessed for years, that is, the appearance of so much whiskey. They never anticipated that they would gaze on such a quantity and be so near to it, and 'yet so far away.'" It was even too much for Judge Partridge, who refused to allow the court clerk to take custody of the liquor and leave it in the building overnight. Instead he ordered, "Take it back to the Customs services or to the Inland Revenue, whose custody it belongs. Bring it back again ... if necessary, but do not leave it here."

As it was, it took several trucks to transport the liquor back to the appraiser's building.

Also admitted as evidence—and objected to by the defence attorneys—were eighty-three one-dollar bills cut in two with liquor orders written on them associated with Mr. Quartararo. As the most active agent of the conspiracy on shore, he was charged with sending them out to the officers of the rum runners to identify his agents for the safe delivery of the liquor.

One of the key points that the American government built their case on was whether *Quadra* was indeed within an hour's steaming of American waters. It was argued that the ship was seized at a distance of 5.7 miles from the Farallon Islands, so a test was made with the launch *C-55*, which was caught with liquor aboard. It revealed that it could traverse 6.6 miles in an hour.

On April 3, 1925, the trial came to a close and the jury presented its judgment. While thirty-two members of the crew of *Quadra* were acquitted, twelve defendants were found guilty. Captain George Ford, First Mate George Harris, Second Mate Joseph Evelyn and Chief Engineer J. H. Mason, along with Vincent Quartararo, Charles H. Belanger and six others were all found guilty on a charge of conspiracy to violate Prohibition laws. Ford received a two-year jail sentence and a ten-thousand-dollar fine; George Harris, thirteen months; Evelyn, ten months; and Mason, a five-hundred-dollar fine. Quartararo and Belanger received two years and a ten-thousand-dollar fine each. Four operators of San Francisco "liquor boat runners" received eight months' imprisonment, and the two others, who pleaded guilty, didn't receive their sentences that day. Later that year, on July 19, *Quadra*'s supercargo, J. A. McLennan, was fined twenty-five thousand dollars.

When *The Daily Colonist* reported on the judgment and sentencing the following day, it noted that the cargo on board was estimated to be worth $500,000 "local price" and, once out on the streets of San Francisco, worth as much as $1,000,000. It also noted that while Captain Ford was now a Vancouver resident, he was actually a Victoria man, as were most of his crew, who had families there. As is happened, many of the shareholders of Consolidated Exporters, as shown in the original indictment, included men from BC's capital city. The liquor export consortium, along with the captain and officers of *Quadra*, appealed the ruling of the United States circuit court and a week later the officers, along with Quartararo and Belanger, who were all imprisoned in Ingleside Jail, were released on bail. The appeal, *Ford et al. v. United States*, questioned Judge Partridge's comment and rulings relative to *Quadra*'s seizure on the high seas. The case was to take some two years before Chief Justice Taft of the US Supreme Court came to a decision with the help of a jury. He overruled their contentions, thereby validating the lower court's judgment. According to the April 12, 1927, *Daily Colonist*, persons aboard vessels may be tried for conspiracy to violate Prohibition laws "when they have entered into an arrangement with persons ashore to accomplish that end, and cannot plead innocence because they had not actually violated the prohibition of customs laws by sending liquor ashore."

Captain Ford, First Mate Harris and Second Mate Evelyn appealed the verdict but in San Francisco on June 22, 1927, Federal Judge Frank H. Kerrigan's court decided in favour of the United States Supreme Court mandate, affirming the decision of the United States circuit court, and the sentences and convictions were confirmed. (Meanwhile, according to British Columbia historian and writer George Nicolson, while all this litigation was going back and forth, apparently a sizeable quantity of *Quadra*'s liquor cargo "evaporated.")

As a consequence, Captain Ford slipped away back to Canada, as did four members of the crew. Unfortunately, by the time the case was over and done with, *Quadra* had deteriorated and rotted away at her anchorage near Government Island, the US Coast Guard's Base 11 in the Oakland Estuary. In March 1929, one of British

Columbia's most historic vessels, was sold at auction for $1,625 and finally in August 1931, depending on the source, she was either scrapped or towed to Los Angeles for use as a fish barge.

Three days following the capture of *Quadra* off the California coast in October 1924, Sherriff H. W. Goggin seized the steamship *Prince Albert* as she lay alongside the Yarrows shipyard dock in Esquimalt, BC, waiting to be overhauled. The Commercial Cable Company had laid two lawsuits against the rum running mother ship through 1924 and 1925 in two separate incidents off California's Rum Row for damaging their Pacific telegraph cable that ran underwater from San Francisco to Honolulu.

One of those who served aboard the *Prince Albert* prior to her being seized was D. O. MacKenzie, a fireman during her three rum running voyages through 1923 and 1924. MacKenzie had initially signed on to the *Prince Albert*—a steel steamer 232 feet in length— as a deckhand while she was registered with Grand Trunk Pacific Coast Steamship Company. At that time, she was a passenger and freight vessel on the Queen Charlotte Islands (today's Haida Gwaii) where, as a young man, MacKenzie was required to learn the ropes: handling general freight, loading cargos of canned salmon and shovelling tons of coal ashore at the canneries.

Although the steamer was registered under Grand Trunk Pacific, Western Freighters purchased the vessel from the Canadian National Railway in September 1923. (A number of Canadian railways that were on the point of bankruptcy had been taken over by the federal government, including the Canadian Northern Railway in 1917 and the Grand Trunk Pacific Railway in 1920. These two western railways, along with some other lines, were consolidated as the Canadian National Railways in 1923.)

Western Freighters Limited, with its head office in downtown Vancouver, was an American-controlled liquor shipping enterprise. As author Philip Metcalfe explained, it was Roy Olmstead, who had been kicked out of the Seattle police force, who came up with the idea of setting up a Canadian-based office. Metcalfe noted that instead of "the quiet respectability that characterized the Canadian hotel men and capitalists behind Consolidated Exporters, Western Freighters was organized by a colourful group of San Francisco bootleggers. Olmstead himself invested $31,000 under the name of 'Steele' and was listed as a director." With its office located across the line in British Columbia, Western Freighters was able to keep *Prince Albert*'s Canadian registry, but more importantly her British flag. According to Metcalfe, Joseph William Hobbs—one of Vancouver's most colourful personalities and a prominent businessman who promoted the city's tallest skyscraper, the Marine Building—was Western Freighters' chief supplier of Canadian liquor. (Hobbs was listed in the 1923 Vancouver city directory as "Manufacturers Agent, representing W & A Gilbey Ltd. (London, Eng.) and Peter Dawson Ltd. Glasgow, Scotland.")

D. O. MacKenzie wrote of his experiences in *The Bulletin: Quarterly Journal of the Maritime Museum of British Columbia* in 1980. MacKenzie soon discovered that *Prince Albert*'s first trip as a mother ship in the fall of 1923 was a far different undertaking from what he had experienced working on her when she was a coastal steamer. They loaded eight thousand cases of liquor aboard in Vancouver harbour and then shifted over to New Westminster where another four thousand cases, along with barrels of bottled beer, were taken on at the Northern Pacific dock. Once her holds were filled, *Prince Albert* steamed down the coast to drop anchor at an agreed-upon location off the Farallon Islands. After a few weeks sitting out on Rum Row, the skipper remarked one day after having trouble raising the anchor that they'd probably got hung up on the Pacific telegraph cable. The skipper would have been aware of the cable and would have had some idea where it had been laid. This was to be the first of two incidents that would come back to haunt them.

MacKenzie noted that generally it was a rather quiet life and, although they were a floating booze emporium, it was the driest

Prince Albert sitting out on Rum Row, Ensenada, in the early 1930s, loaded down to the gunnels with liquor. Fraser Miles collection.

ship he ever worked on. One bottle of beer was given out at lunch time and breaking into the cargo meant instant dismissal. Overall, living conditions on the ship were good and they were fed the best of food since they were kept well supplied by the small fishing craft that came alongside for liquor orders. The larger items such as water and forty-five-gallon drums of fuel oil, along with fresh fruit and other food supplies, were brought out on the American lumber schooner *Point Reyes*. While some of the crew were preoccupied with the task of pumping fresh water and hoisting the drums of oil aboard, MacKenzie said that their other primary endeavour was to open up the hundreds of cases of liquor and repack them in burlap sacks in order to provide better stowage in small boats.

It was on their third trip out during the winter of 1923–24, with the hatches filled with twelve thousand cases of whisky and beer, that MacKenzie described how, again off the Farallones, they came to foul their anchor, this time for certain, on the underwater cable.

He recalled that the weather hadn't improved any so the ship stuck to drifting and steaming back to position, as was so often required of mother ships sitting out on Rum Row. Attempts were made to anchor during calm spells and finally they were successful. Unfortunately, when bad weather returned and the captain called for the anchor to be raised, it got caught up on the Pacific cable.

More than likely, the anchor was dragging a lot due to the vessel constantly having to manoeuvre to stay in position. After the cable was brought up hanging on one of the anchor flukes, a line was tied around the cable and the anchor lowered sufficiently to clear the cable so it could be dropped back into the sea. MacKenzie noted that once free, they returned to their favoured spot a few days later, where the cable layer and repair ship *Restorer* appeared on the scene. (It is interesting to note that no one ever owned up to actually cutting the cable to free their anchor.) Looking back, MacKenzie believed the *Restorer* was there to take photographs to be submitted later in their lawsuit against Western Freighters. (The steel twin screw steamer—which was 358 feet in length and built as a cable vessel in 1903 at the shipyard of Armstrong Whitworth & Company in Newcastle, England—was owned by the Commercial Pacific Cable Company, New York.)

As it happened, Bent Sivertz, who was eighteen years old at the time, signed on *Restorer* in early January 1924 as an able seaman, and was aboard when they arrived off the Farallones later that month to try and locate the damaged Pacific cable and make necessary repairs. Sivertz was assisting the two bowmen, who sat in boatswain's chairs slung from a davit either side of the bow. Here, they were in the perfect position to ascertain the state of the cable as it was pulled up to the surface by a grapnel attached to a line hauled in on the cable winch on the foredeck. Once it was out of the water, they ran along it, examining it for holes or nicks in its armour or insulation that would have caused a short circuit. They weren't seeing any abrasions when all of a sudden, they came up with a bitter end. The cable men had never seen the likes of it; someone had cut the cable clean through with an axe. In his autobiography, *The Life of Bent Gestur Sivertz: A Seaman, a Teacher, and a Worker in the Canadian Arctic*, Sivertz explained, "A ship had hooked the cable on an anchor, hoisted it up to her anchor hawsepipe, and instead of clearing it in seaman-like fashion, had tried to break it." But not only that, after *Restorer* picked up the other end of the cable, they discovered damage all along it with tarred jute torn off, the cable partially flattened and strands of the heavy armour wire broken and wrapped around the cable in loops and whorls. As it appeared to the cable ship crew, *Prince Albert*'s captain had attempted to under-run (pass under) the cable and then put the helm hard over, to jump the cable off. When that didn't work, they grabbed an axe.

Once the repair job was completed, *Restorer* went into San Francisco to lay at the Embarcadero to await orders to make their way to Guam. It was here they learned *Prince Albert* had proceeded into San Francisco the day after the cable went dead. Although it was common knowledge that the vessel was an out and out rum runner, she wasn't arrested once inside American waters. This was due to the fact that she was protected under the "right of innocent passage," considered the first law of the sea, which allows for a vessel to pass through the territorial waters of another state so long as it is not prejudicial to the peace, good order or security of the coastal state.

It was quite common all through the Prohibition years for shipping outbound from Europe with liquor consignments aboard destined for Vancouver to enter San Francisco and possibly discharge other cargo, as well as take on water or fuel, or have repairs made. Still, one is left wondering why the American authorities went easy on the mother ship *Prince Albert*, especially after their somewhat controversial seizure of *Quadra*. Perhaps it was because there were colourful San Francisco bootleggers invested in Roy Olmstead's Western Freighters shipping company and the vessel's cargo. This question would present itself again after the capture of another mother ship, *Coal Harbour*, off the California coast a few months later, when the rivalry between rum running interests, and the extent that they would go to deter their competitors, would come to light in a San Francisco courtroom.

It was while they lay alongside at the Embarcadero that *Restorer*'s crew heard stories from American immigration authorities about the interesting condition they found captain and crew of *Prince Albert* in after they tied up in the Bay City. Apparently, many of them were drunk, while one had to be taken to hospital due to a knife wound. Overall, Bent Sivertz mused that from what he learned about how the ship was being run and especially how the ship's officers would even consider freeing the Pacific cable by under-running it or severing it with an axe, he wondered if the captain was simply incompetent and the chief mate perhaps lacking enough courage to lock him up.

After reading D. O. MacKenzie's account fifty-six years later, Sivertz congratulated him for leaving *Prince Albert* upon arrival back at Vancouver. The deciding incident for MacKenzie and the rest of the engine room staff was the appearance of the US Coast Guard cutter *Shawnee*, which caught them unawares as they all sat on deck enjoying one of the few nice days off the Farallones. The commander of the cutter had taken his ship seawards and then came down on them unnoticed, with its white hull directly in line with the sun. Although the cutter made no attempt to stop them on this occasion (they were probably outside the limit), the encounter put a damper on further adventure for nearly all aboard. MacKenzie said that the prospect of "breaking big rocks into little rocks in a San Francisco jail had no more appeal than taking the chance of being capsized while drifting around the ocean." Sivertz agreed wholeheartedly and commented that "as a seaman during those years, I knew many men who sailed in rum runners. The majority were ruined by it, despite the high pay—or perhaps because of it. Most of us felt it best not to go into the twilight of doubtful legality and fly-by-night work that was both unproductive and boring. Many men engaged in rum running lost quality in character and seamanship."

Back in Canadian waters, *Prince Albert* was put under seizure while tied to the dock in Victoria harbour. On March 5, 1925, the Commercial Pacific Cable Company made an application to the Admiralty court for conversion of the vessel into cash in order to recoup the fines for the two legal claims made against the steamer for damage to the San Francisco–Honolulu cable. The first fine was $99,124 for the first instance that occurred on November 13, 1923, and the second was $91,932 for intentionally severing the cable on January 2, 1924. Apparently the *Prince Albert*'s owners, Western Freighters, were ultimately able to avoid payment by transferring ownership to a dummy company. In 1927, she was owned by Pan American Shipping Company and then in 1930, she was bought by the Atlantic & Pacific Navigation Company, a shipping arm of the Reifel family. As a result, *Prince Albert* ended up seeing the Prohibition years to their end while serving as a mother ship off Rum Row, Ensenada. By 1935, she was back on Canada's West Coast working as a tug with Badwater Towing Company Limited of Vancouver. In March 1945 she towed what was claimed to be the largest Davis raft of logs ever to go up the Fraser River and may have been owned by the Gibson brothers' logging company by this time. By 1949 she was registered under the name of the Tahsis Company of Vancouver, a large lumber mill venture constructed at the head of Tahsis Inlet and owned by the Gibson brothers, recognized pioneers in Vancouver Island's West Coast forest industry.

By early 1925, as more rum running vessels were being seized by the US Coast Guard or lost to marine mishap, there was some concern over the state of the west coast liquor supply fleet and whether it was up to meeting the demands of the export trade. On January 4, 1925, Victoria's *Daily Colonist* reported the unsettling news that "if the report regarding the *Malahat* is confirmed, it will leave but two liquor carriers in operation out of Victoria and Vancouver. These are the *Principio* and *Stadacona*. The *Quadra* is held at San Francisco after being seized by the US cutter *Shawnee* as an alleged rum-runner; the *Prince Albert* is tied up here awaiting the decision on the Commercial Pacific Cable Company's suit for damages, while the *Malahat* is believed sunk." (This report, which originated in Los Angeles, suggested that *Malahat* may have been lost in a bad storm, which was found to be erroneous a few days later.) It would seem then that running liquor by mother ships was in danger of coming to an end, as other news-grabbing US Coast Guard seizures continued to occur through 1925.

Chapter Five

THE DISASTER YEARS PART II

MORE CAPTURES OF CLEAN-CUT YOUNG CANADIAN SEAMEN

The men of Rum Row are principally tugboat men and fishermen who formerly worked in the inland waters about Vancouver. They come south for a year or more at a time, some of them work on the base ships anchored off the coast, and others to handle the fast ex-submarine chasers that carry the liquor to points off the United States where small speedboats come out and unload them. Most of these jobs are entirely within the law—in fact, all of them are except those on the few boats which "run in"—that is, go into American waters to unload. This is very dangerous, for the speedboat operators ashore resent having their lucrative jobs taken away. They revenge themselves by helping the revenue men capture such boats.

—ROBERT DEAN FRISBIE, *RUM ROW: WESTERN*

T hree months after the capture of *Quadra* and seizure of *Prince Albert*, another mother ship, the three-masted auxiliary schooner *Speedway*, met disaster after departing Victoria on January 23, 1925, loaded down with some 17,500 to 18,000 cases of liquor. While her official paperwork stated that the liquor-laden schooner was bound for Champerico, Guatemala, her master, of course, never intended to sail her that far. Instead, a false landing receipt would be written up for the vessel at Champerico, probably around the same time the mother ship was delivering up her cargo far out at sea somewhere off the US coast, or perhaps even before she had left Canadian waters. Hugh Garling noted in one of his *Harbour & Shipping* magazine articles that *Speedway*, which was 155 feet in length, was loaded with a particularly dangerous cargo. Along with the liquor in her holds, she was carrying the equivalent of eighteen thousand cases of distillate (the purest of grain alcohol) in five thousand-gallon tanks below decks and sixty-five drums stowed on deck and in the after-hold. Also, *Speedway*'s auxiliary power happened to be a 250-horsepower gasoline—not diesel—engine. Therefore, apart from the presence of the distillate, her fuel tanks were filled with particularly volatile fuel.

The *Daily Colonist* reported on January 28, 1925, that a Mr. F. M. Bonner of Howe Street, Vancouver, had bought the famous motion picture schooner *Pirate* at an auction sale in Vancouver and changed her to Canadian registry under the name *Speedway*. It was also alleged that Mr. Bonner sold the *Speedway* to a private liquor export concern and the vessel was chartered to Seattle-based Roy Olmstead's Western Freighters, to deliver an order of liquor to Champerico. After casting off from Victoria, *Speedway* cleared Cape Flattery, dropped the towline, and was making good time in the open Pacific with all sail set the following day. (Because of Juan de Fuca Strait's confined waters, large sailing vessels required tugs to tow them in and out of the strait.) Then at three thirty that afternoon, Second Engineer Matthews raised the alarm by shouting, "Fire!! Fire!!" Captain Robert Sinclair rushed down into the engine room with his engineer to discover that the single fire extinguisher was nearly dry of chemical and that the fire pumps couldn't be started. With no means to contain the conflagration, flames were soon racing out of control along deck beams and the deckhead. (The official wreck report stated that the fire was "caused by spontaneous combustion following explosion of gases, in the region of the Standard Exhaust." In other words, the engine backfired.)

Fearing that the flames would soon reach the fuel tanks near the engine room—or even the five thousand gallons of distillate—Captain Sinclair ordered all hands to the boats and abandoned ship. Once two boats were away, they were dismayed to discover that they weren't provisioned with water and decided to return to the ship to secure some, regardless of the risk. The crew was required to wend their way across the water carefully since oil was burning across its surface some fifty feet all around the schooner. By the time they were able to board their ravaged vessel, the fire had burst through the engine room, the cabin was already gone and the masts and sails were immersed in flames.

As they pulled away from *Speedway* around five o'clock that afternoon, there was a thunderous explosion as the distillate tanks blew up, shattering the deck into matchwood and sending steel drums of distillate and cases of whisky flying above the masts 250 to 500 feet in the air. Among the shrapnel falling from the sky was a piece of metal that drove through the bottom of First Mate Metcalfe's boat, causing a good deal of panic before the hole was plugged. After standing off to watch their blazing ship sink beneath the waves, they set sail in a north-easterly direction for the American coast. Vancouver's *Harbour & Shipping* magazine reported in February 1925 that *Speedway* sank about sixty-five miles west of Grays Harbor, Washington.

Captain Sinclair's boat, with six men aboard, wasn't in the water long before it started leaking. Its occupants were preoccupied with bailing as the wind picked up and, by that evening, a heavy sea was running. Early the next morning the Matson Lines steamer *Manulani* was sighted inbound from Honolulu for Seattle. It was good timing since by this time, Sinclair's boat was half full of water that continued to rise regardless of the crew's non-stop bailing. Once they were rescued, the liner dropped them off at the American lightship *Swiftsure* at the entrance to Juan de Fuca Strait. The other boat, with First Mate Metcalfe in charge, was able to safely make its way to Pachena Bay under sail and oar, but what with the heavy seas running that night, the rudder was carried away. Fortunately, despite his exhaustion, Metcalfe was able to jury-rig a rudder, which allowed him to navigate their boat towards land through the rocks and pinnacles strewn along the Vancouver Island shoreline.

The day before all the survivors were to arrive into Victoria harbour aboard the Canadian Pacific Railway steamer *Princess Maquinna*, *The Daily Colonist* ran a large headline across its front page: "Allege Doomed Vessel Hijacked." According to the sensational story, Seattle sources reported that *Speedway*'s liquor cargo had already appeared on the city's "outlaw" whisky market. According to the article, this led to two possible conclusions: either the schooner was hijacked before she sank, or a duplicate shipment of identical liquor (Scotch whisky), "a brand so seldom seen as to be almost unique in this locality has been landed under cover." Of course, once captain and crew stepped ashore in Victoria, the hijacking theory was dismissed.

It was only a week and half later when another disaster at sea involving a Canadian rum runner occurred. This time it involved a Canadian mother ship that went to the rescue of an American vessel in distress. On February 1, 1925, in the midst of one of the worst storms that winter, the 579-ton steam schooner *Caoba*, which was twenty miles south of Tillamook Rock, Oregon, discovered her rudder shaft had cracked. Meanwhile, her seams were opening up from the terrible pounding the vessel was taking and she was rapidly filling with water as the engine room pumps were unable to keep up. To make matters worse, she was at the mercy of wind and wave, having lost steerage. Finally, Captain Wilfred Sandvig ordered his crew of eighteen into the schooner's two boats. They remained adrift for two days before the boat with First Officer A. Rigaula and eight crewmembers in it was picked up by the steam schooner *Anne Hanify*, which passed them over to the Grays Harbor–based tug *Cudahy* to take them into Aberdeen. The other boat, with nine aboard and under the command of Captain Sandvig, was picked up a few hours later off the Willapa Bar by *Pescawha*.

Of Canadian registry, the two-masted 90-foot auxiliary schooner *Pescawha* was built in Liverpool, Nova Scotia, in 1906. Sometime shortly after, she arrived on the West Coast to join the pelagic seal hunting fleet working out of Victoria, but her career in the trade didn't last long. After the fur seal population was nearly hunted to extinction, the United States, Great Britain (representing Canadian interests), Japan and Russia signed the North Pacific Fur Seal Convention on July 7, 1911. It provided for the preservation and protection of fur seals, and all open-water fur seal hunting was outlawed north of the thirtieth parallel. This development had both Canadian and American owners of sealing schooners looking for other endeavours in order to keep their boats working. The perfect opportunity presented itself when the Volstead Act came into effect in 1920. By 1923, *Pescawha* was chartered to Roy Olmstead's Western Freighters and was carrying 1,052 cases of liquor in her holds. The schooner was officially registered in Canada to a James McKinley in Vancouver, who was probably one of Olmstead's Canadian agents that he'd set up across the line in order to facilitate his liquor-importing ventures.

As a mother ship, *Pescawha* was bound for a position off the Columbia River bar when they sighted *Caoba*'s lifeboat and went to the rescue. While retired rum runner Hugh Garling claimed that they were outside United States waters at the time, fellow rum runner and author Fraser Miles argued otherwise. He said *Pescawha* was only six miles off the Washington coast at noon on February 3 when they picked up the survivors and that her captain, Robert Pamphlet, made a grievous error in judgment by not moving his boat farther offshore right away following the rescue. As a consequence, they were still well within American waters when the United States Coast Guard cutter *Algonquin*, which was based out of Astoria, Oregon, the port town at the mouth of the Columbia River, appeared on the scene four hours later. *Algonquin* (*WPG-75*) was 205 feet, 6 inches in length and armed with three four-inch guns,

The schooner *Pescawha*, with a number of Coast Guard men aboard, steaming up the Columbia River following her capture. Washington State Parks and Recreation Commission, #154.1958.506.1.

sixteen three-hundred-pound depth charges, four Colt machine guns, two Lewis machine guns, eighteen Colt .45 pistols and fifteen Springfield rifles.

Once Captain Pamphlet heaved to in order to transfer the rescued crew across to *Algonquin*, a lieutenant from the cutter was sent over and heartily commended the rum runner's officers and crew for their quick action. A headline that appeared in the February 6,

WPG-75 was rated as a cruising cutter, first class, by the US Coast Guard. Here she is on patrol in either the North Pacific or the Bering Sea. Once Prohibition came down, she was reassigned to Astoria, Oregon. US National Archives photo, no. 26-G-05-05-44(11).

1925, *Daily Colonist* summed up what happened next: "Cutter Algonquin Plays Mean Trick on Liquor Carrier" and "Shipping Men Are Up in Arms Over Method Used to Capture Pescawha – Schooner Could Have Stood Well Out to Sea Had She Not Sought to Rescue Part of Caoba's Crew." The situation took a turn for the worse once the lieutenant asked Captain Pamphlet for the ship's papers. When Pamphlet objected, he and his five-man crew were put under arrest while a towline was brought over from *Algonquin* and *Pescawha* was towed into Astoria.

Here, on February 5, 1925, Captain Pamphlet and his crew were brought before United States Commissioner H. K. Zimmerman, who set bail at four thousand dollars for Captain Pamphlet and one thousand dollars for each of his crew, while waiving a preliminary hearing on charges following the seizure of *Pescawha* as a rum runner. Captain Pamphlet responded that he was able to raise bail for himself and his crew, but that he required a few days to do it. At his request, they were all taken to Portland. The *Colonist* went on to say that the slightly more than one thousand dollars' worth of whisky in the *Pescawha*'s cargo that had been removed to *Algonquin* was returned to the schooner, where it was sealed up in the hold.

Three weeks later, the *Portland Telegram* broke the news, "Jury Convicts Pescawha Crew. Ten Defendants Found Guilty on All Counts after Deliberation." (Besides Pamphlet and his five-man crew, the four other defendants were the land-based agents of the operation: Jacob Woitte, Frank M. Raick, Joseph Essex and Tex Smith. These four were up for a number of charges of contravening federal law, the most serious being conspiracy to violate the Tariff Act and the liquor statute.) Captain Pamphlet wasn't impressed with the decision. He couldn't understand why his men were being charged, since they were only acting under orders. He pressed his point by arguing that whether it was a conspiracy to sell Jake Woitte a cargo of whisky or to take washing machines to the Fiji Islands, neither the cook nor the engineer had any say in what they were loading and where it was to be unloaded. Further, "all the men are married, have homes up in Vancouver, and have always been hard working citizens. Most of 'em are veterans of the World War. It's going to be damn hard on them to be locked up for obeying their orders."

It was hoped that the sentencing of Captain Pamphlet and the five crewmembers would be taken up with the British ambassador in Washington and an effort made to get a presidential pardon and a thank you for having rescued all those aboard *Caoba*. While nothing came of this, the citizens of Portland did present Captain Pamphlet—who they had come to hold in high regard—with a gold watch in recognition of his life-saving deed. Still, both captain and his crew were sent to prison upon sentencing. After Captain Pamphlet served out his two-year sentence at McNeil Island federal prison, he returned to Vancouver where he was to pass away two years later.

Fortune also didn't favour his last command, *Pescawha*, after her confiscation by the American government. Outbound on a whaling voyage under her new American owner on February 24, 1933, and crossing the Columbia River bar, she was caught in a bad sou'wester, broke down and was driven onto the north jetty at the mouth of the river, where she was demolished. Philip Metcalfe claimed that the libelling of *Prince Albert* by the Commercial Pacific Cable Company and the capture of *Pescawha* all but ended Western Freighters, driving company director Roy Olmstead one step closer to bankruptcy by early 1925. Of course, the loss of *Speedway* while under charter to Western Freighters at the time didn't help matters.

Two weeks following the seizure of *Pescawha*, another sensational capture took place. This time it was of the three-masted auxiliary schooner *Coal Harbour*, with a large cargo of liquor ostensibly destined for South American ports. She was seized near Bolinas Bay, California, about fifteen miles up the coast from San Francisco, by the US Coast Guard cutter *Cahokia*. At the time of her seizure, she was owned by Archie MacGillis's Canadian-Mexican Shipping Company (which also owned the *Quadra* and the *Malahat*), but was under charter to Consolidated Exporters, which owned the cargo. *Coal Harbour* was 127 feet, 4 inches in length and launched in July 1881 as the lumber schooner *Lottie Carson* from the Hall Brothers shipyard in Port Blakely, Washington.

The *San Francisco Chronicle* claimed that according to waterfront reports, the "Canada Liquor Boat" was nabbed within the twelve-mile limit when she approached Tomales Bay (about twenty miles north of Bolinas Bay) to unload to the small boats of the "mosquito fleet." The big steel tug *Cahokia* was 158 feet in length and like *Shawnee* (the Coast Guard cutter that seized the *Quadra* off San Francisco in October 1924), had a cruising speed of only ten knots and top speed of twelve knots. She was usually based at Eureka, California, eighty miles south of the Oregon line, but was sometimes stationed at San Francisco.

Apparently, *Coal Harbour* or "Gray Phantom" had been standing forty miles outside the Golden Gate for three days after filling

Commander Malcolm F. Willoughby USCGR(T), who wrote the book *Rum War at Sea* back in the early 1960s, claimed that the Coast Guard's seizure of the rum runner *Coal Harbour* was one of the most important in Pacific waters. San Francisco Maritime National Historial Park: B05.19,408 pl (SAFR 21374)

her holds with booze in Vancouver on February 4 and, according to the *Chronicle,* the cutter had been keeping a close watch on the schooner as she skirted the 'danger zone' watching for a chance to slip in and transfer her cargo."

On the night of February 17, while *Coal Harbour* was attempting to unload some of her cargo southwest of the Farallon Islands, *Cahokia* moved in and captured the vessel and arrested its fourteen-member crew. The Coast Guard stated that the rum runner was seized following an hour's chase and claimed that the fugitive vessel even tried to ram them in order to get away.

Once *Coal Harbour* was boarded and the cargo checked, a preliminary estimate by US customs officials suggested there were some ten thousand cases of whisky aboard, which were valued at market prices at more than six hundred thousand dollars, all consigned to John Douglas & Company, based in La Libertad, El Salvador. Two days following this sensational seizure, the *San Francisco Chronicle* reported that, "Whisky consequently shot up ten dollars a case f.o.b. in San Francisco Bay" and that "government officials were elated. They believed last night that they have dealt a death blow to the biggest and oldest of the Canadian smuggling companies—Consolidated Exporters, Ltd. The news caused panic in the bootleg market." Chief Warrant Officer Sigvard B. Johnson, in command of *Cahokia*, also pointed out that there had been a sizeable rum running fleet off their coast just outside the danger zone at the time of the capture. He said that during the chase, two vessels, one a large ship capable of twenty knots, appeared at close range, surveyed the action with their searchlights and then disappeared quickly. Once the Coast Guard had *Coal Harbour* under tow, they were jeered and threatened by the crew of *Malahat*, which the cutter had actually been searching for when it seized *Coal Harbour*. *Malahat* was watching the proceedings and steamed within hailing distance while someone aboard grabbed a megaphone to shout across to *Cahokia*, "She was beyond the twelve-mile limit ... You have no right to take her! You're stirring up a row!" (It is likely that the *Malahat* avoided capture because a single Coast Guard ship was only capable of dealing with one mother ship at time.)

Captain J. M. Moore, commander of the local Coast Guard district, denied the story going around that *Coal Harbour* resisted seizure and the crew even attempted to scuttle their ship. (*Coal Harbour* was found leaking badly in the harbour, what with her being an old wood sealing schooner.) He said that the rum ship's crew had simply refused to catch a towline tossed to them from *Cahokia*, which required Chief Officer C. E. Kipste and three men to run alongside her with a boat and climb aboard to secure it. While *Cahokia* was towing *Coal Harbour* through the Golden Gate, "a swarm of little fishing craft, supposedly headed for the Farallones to unload the rum ship, met them just inside the heads." Government officials were elated with their latest capture. The *Chronicle* article concluded with Collector of Customs W. B. Hamilton stating that "If we haven't gone so far as to break the rum runners' backs, we have at least dealt them about the heaviest blows they can suffer." And once the mother ship's officers and crew were ashore in San Francisco, booked in the city prison and with one hundred thousand dollars in bail posted, Kenneth Gillis, assistant US district attorney, informed the press that a charge of violation of the Prohibition Act would be filed against them all. A reporter with the *San Francisco Chronicle* who happened to be on hand when *Coal Harbour* arrived into San Francisco seemed somewhat surprised to discover that "the crew of the *Coal Harbour* were clean-cut young Canadian seamen, far above the average type of former captured crews."

The *Coal Harbour* was commanded by Captain Charles Hudson, "a grizzled British skipper of 60 years" while his brother, E. W. Hudson, was signed on as supercargo. The ship's complement also included First Mate Eric Best, Second Mate H. Whitmore, Chief Engineer Robert Bell and Second Engineer Frank Gerdy, along with six deckhands, the cook and the mess boy. Bail for the officers was set at ten thousand dollars while those of lesser rank were required to post five thousand dollars. As for the ship's cargo, much of it was piled on her decks in sacks, "ready for quick unloading." Also, it was discovered that most of them contained strips of cork. It was assumed that this was done so that the mother ship could toss the sacks over the bulwarks for her customers to pick up in the event that there were high seas and they couldn't run alongside.

According to the *Chronicle*, "The rum-runners asserted that liquor ship Strathcona [*Stadacona*] stood off the coast for weeks, supplying small boats unmolested." The 168-foot *Stadacona* was originally launched as the steam yacht *Columbia II* in Philadelphia in 1898 and later served with the Canadian Navy in World War I as HMCS *Stadacona*. In 1924 she was owned by Central America Shipping Company of Vancouver but there are no records of what work she did. That same year, the *Stadacona* passed into the hands of Ocean Salvage Company, which was owned (at least on paper) by Joseph W. Hobbs, who converted her into a mother ship. However, it is likely that Ocean Salvage was one of Roy Olmstead's two Canadian liquor export outfits, the second being Western Freighters, for which Hobbs was a Canadian agent. Since the company was nominally owned by Hobbs, however, the vessel kept her Canadian registry.

A few days later, the *Chronicle* also claimed that "the rum-chasers [Coast Guard] have followed the *Prince Albert* [owned by Olmstead] around in circles, but have never taken her" while she was sitting out on Rum Row. The paper went on to say that they couldn't get the "self-appointed investigators" (the crew of *Coal Harbour*) to make a statement as to their intentions upon arrival into the city. The paper believed that the Canadian operators probably "suspect 'double-crossing' and are prepared to expose inner workings of the liquor traffic to get even." Ships and crew working for Consolidated Exporters saw Olmstead's Western Freighters as rivals and figured the company was trying to knock them out of the San Francisco market, especially since investors from the city held an interest in Western Freighters.

It wasn't until December, ten months later, that the *Chronicle* reported that Collector of Customs William B. Hamilton directed a crew of men, under heavy surveillance by customs officers, United States deputy marshals and special agents of the Internal Revenue Bureau, to unload the 9,963 cases of various brands of Canadian and Scotch whisky from the captured boat. "The vessel had lain

This elegant clipper-bowed yacht was originally owned by a New York executive of the Singer Sewing Machine Company. In 1915, she was sold off and commissioned into the Royal Canadian Navy as HMCS *Stadacona*, a depot ship. In 1919, she was transferred to the West Coast and, in 1920, she was paid off to the Minister of Marine and Fisheries as a fisheries protection vessel. Finally, following her career as a mother ship, she was sold in 1928 and went back to being a luxury yacht, the *Lady Stimson*, and when she was sold again, the *Moonlight Maid*. Leonard McCann Archives, LM2018.999.036, Vancouver Maritime Museum.

under seizure off Goat Island [today's Yerba Buena Island] and the collector has been kept in a state of constant apprehension lest rum pirates seize the *Coal Harbour*'s cargo." Once unloaded at Pier 19 down along the Embarcadero waterfront, the schooner was to be towed over to Government Island in the Oakland estuary "to keep company with another captured rum runner, the *Quadra*." And when an unofficial recheck was made of her liquor cargo, around a

Coal Harbour being unloaded of its valuable cargo in San Francisco harbour on December 9, 1925. *Coal Harbour,* December 9, 1925, MOR-0768, in Ships folder, Box Pxs22, San Francisco History Center, San Francisco Public Library.

thousand more cases of fancy imported liquor were found aboard than were listed in her manifest. Deputy Collector of Customs John Toland noted that, the entire cargo was of the finest "stuff," representing twenty-five brands of Scotch and other fine liquors. "There is no question, he declared, but that the whisky has come from Europe and that its owners have evaded duty in Canada in addition to violating the Prohibition laws of this country." It was speculated that this extra stock, estimated to be worth around forty thousand dollars, may have been taken on from another rum runner off the California coast and that perhaps Captain Hudson was planning to do a little trading on his own. Meanwhile, as the *Coal Harbour* was being unloaded ten months after her capture, Captain Hudson and his crew were still awaiting trial and Consolidated Exporters were looking at the possible loss of both vessel and cargo for violating the United States' internal revenue laws.

In a 1967–68 taped interview, Captain Charles H. Hudson recounted to Ron Burton how it was he ended up in command of the mother ship. One day, after he had made a few short trips running rum boats down at the foot of Denman Street in Vancouver's Coal Harbour, he noticed another vessel loading liquor and offered to help. The next day, the boss, Archie MacGillis, phoned to ask why he hadn't bothered to come down and pick up his day's pay. Hudson replied that he just wanted to prove himself. MacGillis consequently offered him the job of running *Coal Harbour*. When he asked how much the pay was, and was informed four hundred a month, he promptly responded, "but if I'm going down to San Francisco for a month it's worth six hundred dollars a month!" His demand was accepted and for the next two years he was kept occupied sailing the schooner, loaded down with over ten thousand cases of liquor, down the coast to sit off the Farallon Islands and wait for American boats to come alongside to pick up their orders. Everything was going along fine until *Cahokia* ran alongside them, banging on their hull, demanding to come aboard.

Captain Hudson noted that he had "an awful lot of fun with it," running liquor down the coast with *Coal Harbour*. He said that they "loaded booze in Vancouver and the boat was so loaded with booze, engine room loaded with booze, house for windlass ... so loaded, six or seven cases high on foredeck." While having a ship filled to the gunnels with liquor would appear to be too much of a temptation for the crew, Hudson spoke to that issue arising on board in his interview. "First trip found crew sneakin' booze, so after that first trip when we got back, fired most of them and employed my own men and from then on never had one scrap of trouble ... I'd put a bottle out on the messroom table after supper, 'there's a bottle of rum there,' and they could come in and have a hot rum. That was the only drinking they did, in three years with the boat they never touched it ... it was extraordinary that you could get your men so you could trust them."

Once *Cahokia*'s towline was on *Coal Harbour* and she was taken into San Francisco, Hudson said they only had to spend one night in jail and were let out on bail for twelve months. "It was sheer heaven, crew all staying in a nice hotel on Market Street!" Still, there was the downside to the adventure for him and his brother. "What's mother going to say to this? Two Hudsons in jail a night ... then out on bail for a year!" As soon as they were set free, and before leaving for home, he snuck back down to the ship and retrieved the logbook that had his entry indicating they were twenty-five miles offshore when captured. When they all returned to San Francisco to fight the case, he was somewhat chagrined to have to sit and watch six lawyers "being paid dollar after dollar after dollar!"

When the trial finally resumed on February 20, 1928, the US district attorney based his case primarily on the navigational fix given by *Cahokia*'s commanding officer in determining the location of *Coal Harbour* at the time of her capture. (The *San Francisco Chronicle* stated that, under the present procedure, the ship was figuratively a defendant before the court, rather than the men who manned it. If the jury found that the schooner was legally seized, Captain Hudson and his crew, along with fourteen others, would be entitled to a trial to determine if they actually conspired to violate the Prohibition law.)

Apparently, when the captain of the *Cahokia*, Chief Boatswain Mate Johnson, asked the Naval Radio Direction Finder Station for a fix by radio, he learned that it put the schooner outside the limit, which didn't agree with where Johnson was saying she was. Later, it was discovered that Johnson subsequently altered the log entry to show the Radio Direction Finder Station's position as agreeing with his own, which he claimed was well within the twelve-mile, one-hour sailing limit.

Federal Judge Kerrigan told the court that evidence could be submitted based on the speed of shore boats ordinarily used to make contact with the mother ships and run the liquor into shore. This ruling was used to force the defence to prove that the steam schooner *Coal Harbour* was beyond the one-hour sailing distance of the fast launches coming out from the beach to rendezvous with her.

Harold Faulkner and James O'Connor, the defence attorneys, continued to press their point that the ship was so far from shore that she was immune from seizure and consequently beyond the court's jurisdiction. In his testimony that first day, *Coal Harbour*'s First Mate, Eric Best, told the court that the captain of the *Cahokia* even yelled across the water when they closed in on *Coal Harbour* that day, "I don't care how far offshore you are!" The liquor ship attorneys would argue in the jury trial in February 1928 that the seizure was made forty-five miles out and that the schooner's log dial even showed that when under tow of *Cahokia*, it was twenty-three miles before she was even abreast of the Farallon light. And, finally, federal attorneys were forced to admit under questioning in court that the rough log of the cutter was missing from government offices.

Captain Hudson refuted all the nonsense that came out in the trial and his indignation comes across in his 1968 interview. "What's the deal here? We know full well our rivals up in Vancouver paid the skipper of that cutter twenty-five thousand dollars to tow us in no matter where we were ... And you know damn well because you've got my copy of the log and that we were twenty-six miles off when they took us in." His lawyer asked, "So what do we do?" And Hudson's remedy? "Give them another twenty-five thousand dollars to commit an un-perjury! ... And they did, and we won our case!" Hudson was fairly certain that the US Coast Guard was ordered to bring *Coal Harbour* in, and probably by the same party who paid the *Cahokia*'s bosun a bribe, in order to create financial trouble for Consolidated Exporters. He was convinced that they hoped it would cost Consolidated Exporters millions of dollars in legal fees and then they would all "throw up their hands and stop rum-running! ... From what I saw during the *Coal Harbour* and *Quadra* seizure incidents, it was proof positive that the States were alive with bribery and corruption. They were saddled with a rotten law they couldn't enforce."

In the chapter devoted to Captain Hudson in *Personality Ships of British Columbia*, Ruth Greene said that First Mate Best, who was "a delightful pipe-smoking Englishman" with an impeccable English public school accent, proved an especially valuable asset in the court case. According to Hudson, it was his first voyage rum running and, fortunately for them, he knew next to nothing about the trade. As a consequence, the prosecutors weren't able to get much out of him other than the rather baffling remark, "Are you threatening me, Sir?" Hudson also recalled that their night in jail had its lighter moments. Best, in his impeccable English, mused through puffs of pipe smoke, "Quite an experience, Sir! What!"

On March 6, Chief Boatswain Sigvard B. Johnson, master of the *Cahokia*, finally admitted on the witness stand that the testimony he had previously given regarding the position of *Coal Harbour* was false. He stepped from the stand to immediate arrest by his superior officers on a charge of perjury and was relieved of his position as commanding officer of Base 11, which was out on Government Island in San Francisco Bay at the time. After deliberating for an hour the next day, the jury declared the seizure of the ship and arrest of the crew an illegal act. Before they retired to come up with this decision, Judge Kerrigan submitted four questions to help the jury determine their verdict. Was the *Coal Harbour* intending to land liquor on United States soil? If so, was she intending to land it from her own decks? If not, was she instead intending to land the liquor by shore boat or fireboat? And finally, what distance could a shore boat similar to those used in such traffic cover in one hour? Of course, the answer to the first question was "yes," the second "no," the third "yes," and the fourth "ten miles." It was the answer to the third that spelled defeat for the government's case. As it was, the defence had argued that seizure was made forty-five miles out and twenty-three miles off the Farallones. In its March 9 issue, the *San Francisco Chronicle* said that they learned from the defence lawyers that this particular headline-grabbing trial was unique in American jurisprudence. "It was the first time, they say, that a jury had been called to decide the question of the court's jurisdiction.

Only questions of facts are for juries. And the primary question of fact (not law) was *Coal Harbour* seized within or without the legal grasp of the United States offshore?"

On June 30, 1928, Federal Judge Kerrigan signed a decree releasing the ship and its $750,000 cargo of liquor. The terms of the decree gave the owners of *Coal Harbour* ninety days in which to check the liquor cargo—barrels of bourbon and cases of Scotch along with a thousand cases of assorted wines, including champagne—and then remove it from American territory. This was of particular interest to the local public, since it had come out during the trial that forty or fifty cases had gone missing at the time the contraband was taken from the schooner and stored in the appraiser's building back in February 1925. Defence Attorney Harold Faulkner replied that this liquor, which had apparently evaporated, was what might be called a "social shortage." The government retorted that rum runners might consider it such. Faulkner responded with a hypothetical question: "What would the Coast Guard consider it since there was a somewhat similar 'social shortage' when the [mother ship] *Federalship* was brought in here a year ago."

At the same time Judge Kerrigan was signing the release for both *Coal Harbour* and cargo to the owners, the Attorney General's office also agreed to forfeit their claim on *Coal Harbour*'s sister ship *Quadra*. Two and half years earlier, after *Coal Harbour*'s liquor cargo was unloaded on the Embarcadero waterfront, she was taken across the Bay to join the other captured mother ship, *Quadra*, where both were left to sit and rot away. In its explanation of the circumstances behind Judge Kerrigan's call for her release in June 1928, the *Chronicle* said, "The *Coal Harbour* will not immediately sail from San Francisco harbour—if she ever sails—for the long anchorage is said to have made her unseaworthy." The final, handwritten entry in the Vancouver Ship Registry for *Coal Harbour* recorded that "Certificate cancelled, and Registry closed this 15th day of October

1928. Vessel sold to foreigners (USA.) Advice of HBM Consul of Los Angeles, Calif. USA."

And as for her cargo? On July 4 of that year, the *San Francisco Chronicle* wrote that "the 10,000 packages of liquor from the rum runner *Coal Harbour* were turned back to the owner by the government yesterday and will be shipped to Antwerp, Belgium, aboard the North German Lloyd liner *Witram*, sailing from this port July 14."

The H. W. McCurdy Marine History of the Pacific Northwest stated that, after several years of lay-up in Los Angeles, *Coal Harbour*, now with "a peculiar bark rig," and under her original name, *Lottie Carson*, went on to appear in several motion pictures, including the 1930s movies *Slave Ship* and *Souls at Sea*, as well as *South of Pago Pago*, which appeared in 1940. Following the entry of the United States into World War II in December 1941, she was re-rigged as a schooner and went on to operate in the Mexican lumber trade.

If one was to believe all the press coverage of the day, it would appear that the US Coast Guard was finally gaining the upper hand with its capture and seizure of a number of rum ships throughout the disaster years of 1924 and 1925, but an article that appeared in a Los Angeles newspaper in May 1925 told a different story. It said that a huge rum fleet, carrying cargos valued at several million dollars, was lying off Southern California. The following summer it was able to report that sixteen Canadian, Belgian, Panamanian and Mexican rum runners, "the greatest mobilization of liquor-laden ships in the history of Pacific rumrunning, are hovering off San Diego." Of course, with all the risks that it entailed, the profits from liquor smuggling were enormous. In 1925, a case out on Rum Row was worth twenty-five dollars; on the beach, forty dollars; to the retail bootlegger ashore, fifty dollars; and to the consumer, seventy dollars, or six dollars a bottle. When one considers that a 1925 dollar would be around fourteen dollars in today's money, the financial rewards were well worth the risk.

Chapter Six

THE MAN WITH THE TRADEMARK SMILE

ROY OLMSTEAD (1926)

"Famous Olmsted [sic] Liquor Case Is Now in Session" declared Victoria's *Daily Colonist* in its January 22, 1926, edition. Across the line in Seattle, it noted, Canadians were to be the first witnesses called upon when presentation of evidence began in federal court against Roy Olmstead and forty-five others who were charged with operating a huge rum running conspiracy. There were actually a total of ninety individuals charged, but the rest either absconded to Canada or became cooperating witnesses, while one didn't show up for court. Western Freighters' Canadian directors, of course, dared not step across the line and come within the court's jurisdiction. In his book on Olmstead, Philip Metcalfe said that, "the courtroom's first two rows were reserved for the 45 defendants, who resembled a crowd at a baseball game: swampers and delivery men in open necked shirts, boat skippers and engineers still tan from last summer's sun. They chewed gum, lolled about with their legs outstretched as if on holiday, or talked confidentially in groups of two or three."

In the government's opening statement, the jury was told that Olmstead and some of the other defendants owned stock in two liquor export houses in Canada and possessed three vessels, the *Prince Albert*, the *Pescawha* and the *Coal Harbour* (the last of these was owned by Archie MacGillis's Canadian-Mexican Shipping

The former halibut boat *Chief Skugaid* alongside and loading from the mother ship *Lillehorn*. Fraser Miles collection.

Company and most likely being chartered to Western Freighters). The federal court alleged that the vessels carried liquor manifested for points in Central America and South America, but transferred their cargo in Canadian waters near Victoria to other boats, which again shifted it to boats making for Seattle, Tacoma and other places in Puget Sound. (While all three vessels may have transferred liquor off the entrance to Juan de Fuca Strait, they were primarily operating as mother ships delivering up their cargos off the coasts of Oregon and California.) The paper said that representatives of the Victoria and Vancouver customs houses, along with two officials in the Preventive Department, were present and were probably going to be called to give evidence.

Western Freighters was the liquor export shipping company that Seattle-based Olmstead set up in British Columbia and whose fleet included the above-mentioned steamer *Prince Albert* (capacity about 20,000 cases) and the schooner *Pescawha*. Not included in the indictment were his other vessels: the steamer *Stadacona* (capacity about 22,500 cases), the schooner *Speedway* (capacity about 24,000 cases) and halibut boat *Chief Skugaid* (capacity unknown).

Roy Olmstead had arrived in Seattle from Nebraska as a nineteen-year-old with his family in 1904. He began working in the Moran Brothers Company shipyard as a metal fitter and in 1910 when there was a slowdown in the industry, joined the Seattle Police Department with one of his brothers as temporary policemen. Described by Philip Metcalfe as "tall and competent, with a quick intelligence," Olmstead rose quickly in the ranks with his penchant for hard work, easygoing self-assurance and ability to make quick and sure judgment calls. He was soon recognized by his colleagues and superiors as a born leader. In 1910 he was promoted to desk sergeant in the booking office and then in 1916 was appointed acting lieutenant, which, at thirty years of age, made him the youngest lieutenant on the force at the time. Another factor that probably helped to further his career was that the personable Olmstead "dressed well and employed the vocabulary of a man whose mother was a college graduate."

Olmstead's promotion occurred just after State Initiative Measure No. 3, prohibiting the manufacture of alcoholic beverages and their sale in bars and saloons, was enacted on January 1, 1916. Although he was never assigned to the Seattle Police Dry Squad, Olmstead did take part in a number of raids and arrests of bootleggers and got to see for himself how difficult it was to enforce the new law. More importantly, he also witnessed firsthand the vicious competition between two rival gangs whose attempts to dominate the trade led to open warfare on the streets. It didn't take long for him to recognize that the outright anarchy and lack of organization in these gangs was primarily responsible for the dead and wounded on the streets of the Queen City. And he couldn't help but notice that

According to biographer Philip Metcalfe, Roy Olmstead, pictured with his wife, Elise, dressed well and employed the vocabulary of a man whose mother was a college graduate. Museum of History & Industry, *Seattle Post-Intelligencer* collection, 1986.5G.2261.1.

there was one thing bootleggers weren't lacking for: money. To the young police officer, it was an economic opportunity that just couldn't be ignored.

Olmstead was quick to realize the bigger money was to be made from rum running if it was managed more efficiently by someone with basic administrative skills and good business sense. The formula was simple: British Columbia, just across the line, was readily accessible, with an unlimited source of quality liquor, while Seattle provided an excellent market for the product. As well as being a rather imposing physical figure, especially when dressed in his police uniform, Olmstead also possessed a very charming personality, enjoyed socializing and was well respected throughout the local community for his intelligence and initiative. Even though he was the youngest lieutenant in the Seattle Police Department, where he was referred to as the "baby lieutenant," he realized he was never going to get rich working as a cop, so he set out to improve his prospects. He went after the quickest, easiest money he could find: bootlegging and rum running.

Once the National Prohibition Act went into effect, Olmstead wasted no time establishing himself in the business and was well on his way to becoming Seattle's leading importer of illicit liquor, despite the fact that he was still a member of the police force.

Olmstead started out modestly by taking an interest in the Shipyard Service Station, which was selling oils, gasoline and accessories down by the Seattle docks. It was an ideal location, since little attention was paid to trucks hauling heavy loads late into the night. When Olmstead and his gang (which also included Seattle Police Sergeant T. J. Clark) were caught by US Prohibition Bureau agents while unloading Canadian whisky from a launch at Meadowdale three miles north of Edmonds, Washington, on March 22, 1920, it appeared that Olmstead's foray into rum running was doomed. But Olmstead escaped the roadblock by jumping in his car and driving around through the bush. Unfortunately, the federal agent in charge notified the Seattle Police Department of the capture of one of their sergeants and said he also recognized Lieutenant Olmstead at the scene. Olmstead was ordered to surrender. Besides the two police officers and nine bootleggers, six automobiles and one hundred cases of whisky were netted in the trap.

Olmstead, of course, was summarily discharged from the force and indicted by a federal jury where he pleaded guilty and was fined five hundred dollars. While on the surface it appeared that the enforcement of Prohibition in Washington state was off to a good start, it had an unintended consequence. Once he was cashiered from the police force, Olmstead was free to devote all his time and energy to pursuing his more lucrative profession and he was soon running more liquor into Washington than any of his competitors. After the trial, when he found himself besieged by offers from prominent citizens of Seattle eager to tap an unlimited source of liquor in BC, the once highly respected police officer was well on his way to building an empire that was to rival Al Capone's Chicago gang in the scale of its activities.

But unlike Capone, Olmstead abhorred violence and forbade his boats and their crews from carrying firearms. Instead, he relied on fast boats with operators well versed in evasive manoeuvres to avoid any confrontations with the US Coast Guard, rivals or gangster-type hijackers. Still, Johnny Schnarr pointed out that Olmstead wasn't above resorting to strong-arm tactics if need be. He described a meeting with Olmstead's rival, Pete Marinoff, in a Tacoma coffee shop in the 1920s, when Marinoff described an incident that occurred while one of his boats was taking on fuel down at a Seattle dock. Someone pushed a forty-five-gallon drum of oil off the dock onto the boat fifteen feet below, and while it missed the fellow fuelling up, it did make a sizeable hole in the deck. (Marinoff wasn't able to determine who performed the deed, but he was fairly certain who gave the orders.)

In a story he wrote on West Coast rum running that appeared in the first edition of *Raincoast Chronicles* back in 1973, Ed Starkins said, "Olmstead's methods for bringing contraband liquor into the country had the sophistication and intelligence of the masterminds in a James Bond novel." The complex business operation he managed to put together comprised a small army of dispatchers, salesmen, bookkeepers, warehousemen, mechanics, drivers, boat crews and legal counsel, and was soon one of Puget Sound's largest employers. Along with a fleet of chartered vessels, there were the numerous cars and trucks, as well as a farm where his liquor was cached. Still, the key to his success was his in with the Seattle police force. He not only bribed local law enforcement agents but was also clever and convincing enough to bribe local members of the US Coast Guard and other federal officials who would have been unreachable to shadier characters involved in the trade.

But just as important was Olmstead's personal charm, which, when combined with his skillful persuasive abilities, allowed him to develop the right connections in the Queen City. As Hugh Garling wrote in a *Harbour & Shipping* article, "To the many citizens of Washington, for whom the Eighteenth Amendment was hateful and repulsive, Roy Olmstead was an altruist of the first order. He was welcomed by the exclusive Arctic Club, and many other prominent citizens, including William Boeing—the airplane manufacturer, who was not only a customer but a good friend—public officials, professional men, merchants and bankers. He was the toast at parties, and it would give a person a sense of importance to be able to say, 'As Roy Olmstead was telling me today ...'"

Still, it was two inventions, the automobile and telephone, which had transformed life and Western society dramatically by the early twentieth century, that Olmstead exploited to his full advantage. Bootlegging at its basic level was a straightforward delivery system. Customers phoned in their orders and after enough were collected, a "bottleman" would drive around the city to make deliveries of bonded whisky, rum, brandy or gin (all individually wrapped in manila paper) to various residences or places of business. (The trucks used for deliveries were marked "Oriental Bread," "Fresh Meats," "Dairy Products" and "Cookies.") The whole rum running and bootlegging operation was so well organized and managed that at its peak in 1924, Olmstead's operation was delivering two hundred cases throughout the Seattle area daily while grossing from two hundred to two hundred and fifty thousand dollars monthly.

With his substantial earnings, Olmstead was able to buy himself and his fiancée Elise Campbell—an attractive English girl he'd met in Vancouver—an elegant mansion in the exclusive Mount Baker district overlooking Lake Washington. Here, once again, he took advantage of the latest technology after meeting a keen young inventor, Al Hubbard, who was building a thousand-watt broadcasting station. Olmstead invited him to move into his home, where Hubbard installed the radio transmitter in an upstairs bedroom. They established the American Radio and Telephone Company as one of Seattle's first commercial broadcasting studios, KFQX, which later became the radio station KOMO, which is still in existence today.

Once KFQX went on the air in October 1924, it began broadcasting at dinner hour with news, weather and stock reports. While the station featured popular music of the day later in the evening, it was Elise Olmstead reading children's bedtime stories that really

Elise Olmstead came up with the idea of installing a 1000-watt broadcasting station upstairs in their Mount Baker home. Set up as radio station KFQX, it was one of the most powerful stations in the nation. Museum of History & Industry, 1987.30.1.

drew listeners. These broadcasts were also of particular interest to Prohibition agents, who were convinced that the adventures of Winnie-the-Pooh, Flopsy, Mopsy and Cotton-tail were being used as a cover to send coded messages out to Olmstead's boats inbound with Canadian booze and let them know whether the landing areas were safe and free of agents.

The efforts to bring down Roy Olmstead and his organization by Roy C. Lyle, Prohibition Bureau director for Washington state, and William Whitney, assistant director and legal advisor, began to pay off by the summer of 1924. The first big break came when *Eva B* was captured by Captain A. R. Bittancourt of Canadian customs and his crew in *Winamac* in October 1924 and the rum running launch's crew implicated Roy Olmstead under interrogation. Still, it was the land-based investigation carried out by Prohibition agents that allowed them to move in and try to shut down Olmstead for good.

After a large body of evidence was compiled following an extensive surveillance operation that included informants and hundreds of hours of wiretap information collected from Olmstead's operatives' telephones, Prohibition agents obtained a search warrant and raided Olmstead's "snow-white palace," as the couple affectionately called their Mount Baker home, on November 17, 1924. Two months later, on January 12, a federal grand jury convened to hear evidence in its final report and Roy Olmstead and ninety associates—a motley collection of telephone dispatchers, swampers (deliverymen) and boat crews—were indicted on two separate conspiracy charges. Also among those charged were a number of Canadians, including Russell Whitehead, president of National Canners Limited of Vancouver; F. R. Anderson, a Victoria lawyer; along with twenty-seven directors of Consolidated Exporters and the entire crew of the Canadian mother ship *Quadra*. Twenty of the Americans charged subsequently absconded to Canada, while Western Freighters' Canadian directors decided it was best to remain north of the border. In the end, the US Marshal was only successful in capturing fifty-nine on their list.

The first indictment was for the illegal importation of liquor from Canada and the second for its sale in Seattle. Since the prosecution based its case on some seven hundred pages of wiretap evidence, the trial became known as the "Whispering Wires Case" and soon laid claim to being the largest trial for violation of the Eighteenth Amendment, the Volstead Act, in the United States of America. Six years of intensive Prohibition enforcement, both on land and at sea, appeared to be finally paying off along the West Coast that year. The US Coast Guard had managed to capture and seize three mother ships—*Quadra* in October 1924 and *Pescawha* and *Coal Harbour* in February 1925—and their trials were also proceeding through the courts. Now optimism reached a high-water mark with the Whispering Wires Case in Seattle. Still, prosecution was postponed for nearly a year after a return of the indictments.

Then early on the Monday morning of January 18, 1926, and into the seventh year of US Prohibition, spectators began lining up outside the courtroom on the fourth floor of the Federal Building in Seattle. As Olmstead biographer Philip Metcalfe pointed out, the prosecution of the Whispering Wires Case was to be one of the longest trials in Prohibition history.

Finally, on February 20, 1926, the federal grand jury returned its verdict. While a number of those brought to trial were acquitted, Olmstead and his attorney, Jerry Finch, along with a number of ringleaders, boat operators and drivers, were found guilty on both counts. Olmstead was sentenced to two years in McNeil Island federal penitentiary and fined $8,000 for Count I, conspiracy to import liquor, plus he was given another two years and $1,613.50 for Count II, conspiracy to transport and deliver the product. In passing sentence on Olmstead, Judge Jeremiah Neterer commented that "the damage ... to organized society and to the government of this country is incalculable. It was amazingly wide spread, but a short step to the undermining of those institutions which are so sacred to us ... As to you, Roy Olmstead, I'll say this ... if the same constructed force and organizing ability which was devoted to this enterprise had been used legitimately, in harmony with the law, the final result would have been marvelous ... If you don't know you have erred grievously, I am sorry for you."

Burdened with a massive legal bill, Olmstead was forced to sell his Mount Baker home. His conviction and sentencing didn't seem to dampen his active involvement in the liquor trade to any degree and he continued smuggling BC booze into Washington and Oregon states, both by sea and land, while appealing his conviction. (He was in and out of McNeil Island a few times during his appeals to higher courts.) A report prepared in December 1927 by Special Agent in Charge Ralph H. Read, San Francisco, addressed to the commissioner of Prohibition, Washington, DC, and directed to the attention of its field division, proved particularly revealing that government officials were being kept well informed of what shenanigans Roy Olmstead and his business associates were still up to:

> Our preliminary investigation discloses that the city of Portland, Oregon, is being used by this ring of conspirators as a distributing centre, the liquor being transported not only to Portland, but also to Tacoma and Seattle. Mr. Herrick has reasonable grounds for believing that the Sterling Traders Ltd., of Vancouver, BC, is concerned with the sale of the smuggled liquor, and the shipment of it by sea to points off the Oregon coast, and the liquor itself is probably being purchased by Roy Olmsted [sic], Arthur Boyd, Ed Morris, all of Seattle, Washington, and suspended Prohibition agent Alfred Hubbard ... The United States Attorney for the district of Oregon considers this the largest and most important smuggling case that has come up in his district during his term of his office.

In the end though, the Prohibition service's agents in Portland weren't required to proceed with their investigation into the ongoing smuggling conspiracy much further. Olmstead was taken out of circulation seven months later, after US Marshals picked him up and returned him to McNeil Island to complete his original four-year sentence on June 28, 1928. Olmstead's appeal had been based on one argument: that evidence obtained by wiretap was unconstitutional and that an individual's right to privacy, and particularly against self-incrimination, was protected by the US Constitution. But the US Supreme Court decided otherwise and in a majority decision upheld his conviction, thereby affirming the government's right to utilize wiretaps in criminal investigations. Of course, while he was serving out his sentence in prison, it didn't take very long for other bootleggers to fill the very large gap he'd left in the local liquor trade. Finally, "the man with the trademark smile," Roy Olmstead, was released in May 1931. He was picked up by his wife, Elise, at the prison dock and the couple drove home to Seattle, very keen to set out on a new path. After his conversion to the Christian Science faith while serving time in prison, Olmstead returned to civilian life to put his considerable talents and people skills to better use in a legitimate trade: selling furniture.

Chapter Seven

THE COAST GUARD YEARS

GRANDDAD AND THE US COAST GUARD TAKE CHARGE (1926-29)

I t didn't take long after the enactment of the Volstead Act before a sizeable portion of the American public began brewing liquor secretly in their homes, while hidden away out in the backcountry, thousands of stills were set up by moonshiners. Regardless, a large part of government resources was dedicated to putting a stop to the better quality rum, gin, whisky and beer—particularly those bearing labels of the established distilleries and breweries in Europe—from flooding across the border from their neighbour to the north. During the first year of US Prohibition alone, it was estimated that the annual value of Scotch whisky imported into Canada increased from $5.5 million to $23 million. One of the greatest challenges during the Prohibition years was trying to control this massive inflow of liquor arriving into the US by sea in everything and anything that could float. It became the greatest challenge to Prohibition authorities and the responsibility for trying to curtail it was left primarily to the United States Coast Guard.

The United States government left the burden of the overwhelming administrative responsibilities called for in trying to enforce the Volstead Act with the US Treasury Department's Bureau of Internal Revenue, which created a separate Prohibition Unit in late 1919 to oversee the war on the illegal liquor trade. The various subdivisions within the unit were put in charge of manpower and resources on both land and sea. Working alongside the Prohibition agents and other divisions within the Bureau were the US Customs Service and US Coast Guard. These federal government resources were also supported and bolstered by state, county and municipal officials, along with their sheriff departments and police forces.

For the US Coast Guard, enforcement was a particularly daunting task throughout the early years of Prohibition. As Commander Malcolm F. Willoughby noted in his 1964 book, *Rum War at Sea*, a detailed account of the service's record during the Prohibition years, "The smuggling of liquor from the sea began in a small way but grew to immense proportions. At first, the enforcement of the prohibition law fell chiefly upon the customs and prohibition agents ... Prohibition agents could intercept liquor being landed on the wharfs of New York and other large coastal cities to some extent, and in some ports they had boats of their own to catch the small craft operating in local waters. The task, proved far beyond their capabilities ... It soon fell squarely upon the US Coast Guard to suppress smuggling from the sea ... Even with enforcement efforts at their best, it was like trying to stem a flood with a rake. And the flood started within two weeks of the advent of prohibition."

The Coast Guard found itself simply overwhelmed with the resourcefulness of rum runners and was somewhat chagrined that it was now charged with enforcing the Volstead Act at sea. It didn't take long to discover that this added responsibility was a considerable drain on its resources and, most importantly, interfered with its principal peacetime duties. These were to protect life and property at sea, maintain the country's aids to navigation, including lighthouses, enforce maritime, tariff and trade laws and prevent smuggling. On top of all this, the Coast Guard was required to continue to maintain military preparedness. (In contrast to the Canadian Coast Guard, a special operating agency within Fisheries and Oceans Canada, the United States Coast Guard has always been both a federal law enforcement agency and a branch of the US Armed Forces. Indeed, until the Department of the Navy was established by Congress in 1798, the Coast Guard served as the nation's only armed force afloat.)

As Commander Willoughby explained it, the responsibility for patrols of Juan de Fuca Strait, the Strait of Georgia, and the approaches to Puget Sound was all left in the hands of the commanding officer of the Port Townsend base. There were subsidiary bases at Anacortes and Port Angeles but most of the Coast Guard's meagre resources were dedicated to the southern shore of Juan de Fuca Strait from Port Angeles to Neah Bay, in order to deter landings by water. From 1920 through 1924, the US Coast Guard was ill-equipped and severely challenged by the lack of suitable vessels in its attempt to curtail the flourishing trade in booze. The big cutters like the 240-foot *Haida*, the 205-foot, 6-inch *Algonquin* and the 198-foot *Bear* spent most of the year based in Alaskan and Bering Sea waters on patrol, but *Algonquin* was sent south in the off-season and *Haida* spent some time patrolling off the Washington coast. ("Cutter" is the term used for all US Coast Guard commissioned vessels 65 feet in length or larger. They were classified under three categories: cruising cutters capable of extensive sea-going operations, inshore patrol vessels which were often US Navy subchasers and World War I veterans, and the harbour cutters, most of which, like *Arcata*, as old rum runner Hugh Garling noted, "were simply tugboats utilized for routine harbour duties and particularly unfit for zealous booze smugglers.")

Enforcement work throughout Puget Sound was left primarily to three vessels: the 67-foot harbour cutter *Guard*, which was only capable of a maximum speed of ten knots with its oil-fired triple-expansion steam engine; the 61-foot *Scout*, which carried a crew of six and was powered by a gasoline engine; and the 85-foot wood-hulled revenue cutter *Arcata*, a steam tug built in Oakland, California, in 1903 by W. A. Boole & Sons. *Arcata* was purpose-built for the US Revenue Service to carry compliance officers around the San Francisco Bay Area. But after only one year of service, *Arcata*,

In 1936, the thirty-three-year-old cutter *Arcata* was declared surplus and sold to the Foss Launch and Tug Company, where she was renamed *Patricia Foss*. The tug was finally laid up for disposal in 1963. On the night of July 4, 1967, she was set ablaze in Tacoma's Commencement Bay as part of the Independence Day celebration. us National Archives, photo no. 26-G-116-F-1.

with her crew of fifteen, was transferred to Puget Sound and stationed at Port Townsend. In 1926, she was shifted over and stationed out of Seattle.

Johnny Schnarr, who dedicated his rum running career to the highly risky and dangerous business of jumpin' the line directly across the Canadian border into Washington state waters, noted that *Arcata* seldom ventured as far north as Haro Strait, where much of the trade transpired, and instead concentrated on patrolling the waters around the mouth of Puget Sound. Driven by an ancient "coffee grinder" steam engine, *Arcata* was good for twelve knots at the most and could only stand by and watch helplessly while out on patrol as rum runners, who favoured fast hulls and powerful engines, sped by through the San Juan Islands and Admiralty Inlet, destined for the remote islands scattered throughout Haro Strait. (The fast runners were referred to as "blacks" by the Coast Guard since they were always painted that colour in order to make it next to impossible to spot them out on the water at night.)

Yet the old steam tug did receive much-deserved recognition for the number of captures she was able to achieve, primarily due to the skills of her captain, Chief Boatswain Lorenz A. Lonsdale. Hugh Garling noted that the five-foot-tall Lonsdale earned an outstanding record primarily due to his wealth of local knowledge and cool determination,

Boatswain Lorenz Lonsdale, Master of the *Arcata*, was a strict teetotaler and stern disciplinarian. He was held in high regard by both rum runners and Coast Guard alike who affectionately referred to him as "Granddad." Puget Sound Maritime Historical Society

combined with his greatest asset: patience. More importantly, he was a strict teetotaler and stern disciplinarian who refused to be corrupted. As a consequence, he was held in high regard by both rum runners and guardsmen alike, who affectionately referred to him as "Granddad."

Arcata's first capture occurred in 1920, just after Prohibition went into effect, and many bootleggers hadn't yet caught on to the fact that the US Coast Guard was now receiving direction from the Prohibition Unit. (Bootlegging was already well established since Washington state had just gone through four years of state Prohibition before it was brought in nationwide.) The launch

George F was caught by surprise by this development as, with decks piled high with Canadian liquor, she cruised into Port Townsend past *Arcata* and tied up at the dock. Consequently, when the rum runner came alongside the dock, guardsmen were right there to grab the lines and along with them, the boat and its cargo.

One trick Granddad relied on was to take advantage of any fog that happened to roll in. One afternoon as *Arcata* was lying off Marrowstone Point at the entrance to Admiralty Inlet, they watched helplessly as the fast gasoline launch *Searchlight* cruised by, her skipper giving a friendly wave as they headed out of Puget Sound to pick up their next order. Later that day *Arcata* anchored inside Foulweather Bluff, near the north end of the Kitsap Peninsula, well hidden by thick fog. Just before daybreak as the fog began to lift, they spotted *Searchlight*, now sitting at anchor only a few hundred yards away. A boat was sent over from the cutter, and the crew rowed alongside the rum runner and climbed aboard to stir the crew awake with the barrels of their .45s. But after a search uncovered only two bottles of beer, Granddad ordered the boat to the beach. Here they came across some freshly disturbed sand, so they started digging to unearth twenty-four cases of liquor. After *Searchlight*'s crew 'fessed up to the cache being theirs, *Searchlight* was seized.

Fast rum boats headed north that happened to speed by *Arcata* often took great pleasure in shouting obscenities across the water. But two of Olmstead's top boat operators, Clint Atwood and Prosper Graignic, remained courteous and were always friendly. Graignic, in his boat *Three Deuces* (which escaped capture by Captain A. R. Bittancourt in *Winamac* back in October 1924), would even come alongside the cutter to swap tobacco and try to fish information out of captain and crew about what the Coast Guard was up to. Captain Adrian L. Lonsdale, the grandson of Lorenz A. Lonsdale and the commanding officer of the icebreaker USCGC *Edisto*, described the fast launch in a story titled "Rumrunners on Puget Sound" that appeared in *American West* magazine in 1972. *Three Deuces*, powered by a powerful Liberty engine, "was thirty-eight feet long and had austere accommodations. Graignic piloted the craft from a small cockpit forward; behind him was an open cockpit covered by canvas peaked with a ridge pole, and behind that was the engine and another small compartment. The boat could carry two hundred and fifty sacks."

Chief Boatswain Lorenz A. Lonsdale was not only frustrated with the speedy and powerful boats that they had to contend with, but his crew was only armed with Springfield rifles to fire on the runners when they ignored orders to stop. In one instance, *Arcata* spotted the rum runner *June*, a new boat capable of twenty-five knots, headed south through Admiralty Inlet and only two hundred yards off. When Lonsdale signalled her to stop, the operator responded by ramming his throttle wide open and speeding off. In frustration, Granddad reached for one of the rifles and started shooting but it didn't have any effect. After receiving a report of the incident, the Coast Guard tug *Scout* came across *June* tied to the dock in Lake Union, after having successfully delivered her booze cargo but at a cost of seven bullet holes in her hull.

After a number of encounters where rum runners simply ignored the Coast Guard to race away to safety, Lonsdale finally demanded a semi-automatic one-pounder deck gun be mounted on *Arcata*, capable of firing rounds four thousand yards and proven to be particularly accurate to around five hundred. The beefed-up armament aboard soon proved well worth the investment. In June 1924, a Victoria paper reported that four men were seriously burned and three wounded by bullets when their boat's gas tank was pierced by the cutter *Arcata*, which they came upon in Mutiny Bay, eighteen miles north of Seattle in Admiralty Inlet. "The 400-gallon tank of the pursued craft blew up when a shot from the Arcata's one-pound gun went through it, and the motorboat, which was 36 feet long and speedy, took fire ... Ralph O'Leary, the least injured of the four persons aboard the motor boat when the *Arcata* appeared five hundred feet away, said that the party was out on a hunting trip, but was running without lights." Once *Arcata* opened up with her one-pounder and rifles, O'Leary was hit by a bullet while at the wheel and two of his crew jumped into Mutiny Bay when the tank blew up. Guardsmen were able to get aboard and search the burning vessel, where they found a quantity of Canadian liquor. Meanwhile O'Leary, who was wounded in the leg, manhandled unconscious crewman Ernie Jackson, who had been hit in the head, aboard the cutter, which also retrieved the two men in the water.

Hugh Garling claimed it was Olmstead's chief competitor, Pete Marinoff, who had been fired upon and arrested by Granddad more than any other rum runner. In his biography, Johnny Schnarr said, "'Pete Marinoff was short and dark and spoke with a bit of an eastern European accent. He owned a 'near-beer' brewery in Olympia, but made his real money bootlegging the good liquor we were bringing into Washington.'" Like Schnarr, Marinoff had a passion for fast boats and, at one point, owned a fleet of seven that he kept well occupied making runs from the Discovery and Chatham Islands and East Point on Saturna Island down to Seattle and Tacoma.

One morning, probably in 1924, while anchored at Allan Island, Marinoff was woken by *Arcata*'s dinghy bumping alongside. Marinoff was aboard *Alice*, 50 feet in length and housed over and fitted with a mast and boom. After a thorough search, nothing was to be found, and Granddad reluctantly had to inform Marinoff that he was free to continue on. But the Coast Guard became suspicious when he didn't seem all that keen to be on his way, and Guardsmen were sent ashore, where sixty-five cases of liquor were found. *Arcata* returned to *Alice* and this time a full safety check of equipment was undertaken. As a result, when Marinoff couldn't produce any fire extinguishers, and his life jackets were found to be in poor condition, he was taken in for violation of safety equipment regulations.

At Seattle, when *Arcata* tried to bring *Alice* alongside, the tow rope got fouled on *Alice*'s prop. As the crew fished around under the

After the *Alice* of Seattle was captured inside Upright Head, Lopez Island, in the San Juan Island Group by *CG 2354*, 114 cases of liquor were discovered concealed in two compartments under her hull. Bower & Brady photo courtesy of United States Coast Guard Museum Northwest, Seattle boat with a pike pole trying to free it, they happened to hook onto an iron piece formed into a U shape with cords hanging from it. When one of the cords was pulled to the surface, a sack was found hanging on the end. After twenty-four sacks of liquor were pulled up by their cords, the vessel was beached, where it was discovered that the sacks had been stuffed in compartments built on each side of her hull along both sides of the keel. The sacks were placed in the opening of the compartment at the after end and shoved forward with a twenty-two-foot pike pool. This was probably all accomplished in a boathouse or when the vessel was beached at low water. With the compartments loaded, a thin board was nailed over the opening. Now with his subterfuge exposed, Marinoff lost his boat, which was subsequently auctioned off.

Less than a year later, as Lonsdale and crew watched the steamer *Mukilteo* passing by southbound in Admiralty Inlet, they noticed something peculiar. She appeared to be stacked with too many deckhouses. They drew closer to see what this was all about. It was Marinoff's former launch, *Alice*, now numbered M-817. She was found trying to sneak by while concealed on the other side of the steamer. Once *Arcata* opened up on the rum runner, her crew jumped to it and started heaving sacks overboard. But when *Arcata* caught up with them and no liquor was found aboard, she was allowed to continue on her way. Captain Adrian L. Lonsdale claimed that it was later reported that this particular liquor cargo was hijacked from the Canadian boat *Pauline*. (This was not the same *Alice* owned by the notorious Egger brothers. Hugh Garling described the Eggers' *Alice* as a powerful, black-hulled speedster, a "black," not a "grimy ex-Navy motor sailer," as Adrian Lonsdale described the M-817.)

In spite of the cool and professional demeanour that Lonsdale projected to those working on both sides of the law, he found his Coast Guard endeavours throughout the years of the Noble Experiment discouraging. He did succeed in making a number of arrests but, because the courts were so overwhelmed with Prohibition cases, most of the suspects ended up free on bail pending trial. As his grandson Captain A. L. Lonsdale explained, "In the meantime, he would get back to his trade, and eventually have three or four trials pending. As boats that were confiscated were laid up, political pressure would mount for their public auction. Eventually the US marshal would submit to the pressure, hold the auction, and soon the Coast Guard was back out chasing the same boats again." Although both the Coast Guard and Prohibition agency were opposed to this system, they did benefit themselves at times by purchasing the faster boats for Coast Guard use.

Despite its best efforts, the US Coast Guard was still severely challenged by a lack of adequate resources—both vessels and manpower—to adequately cope with its daunting enforcement challenges. In the early years of Prohibition, the service estimated that it was only able to intercept a meagre 5 percent of the liquor entering the US by sea. The Treasury Department continued to pressure Washington to come up with the funding to at least double the strength of the Coast Guard and establish a chain of boats along both coasts. Rear Admiral F. C. Billard, Coast Guard commandant, had asked for $28,500,000 to build a fleet capable of taking on the sea-going component of the rum war. If the Coast Guard were supplied these funds, they would be able to build 20 new cutters, 200 cabin cruisers and 91 smaller motorboats, which would be operated with around 3,535 additional officers and enlisted men. The *New York Times* headlined the news by stating that now the nation's Coast Guard was "declaring open warfare on rumrunners and dope smugglers."

Commandant Billard announced, "I have taken the position here in Washington that the Coast Guard has never fallen down in any task assigned to it, and that we will stop rumrunning on our coasts if given the funds for which we asked. For the honour and prestige of the service, we must not fail and we will not fail."

Finally, in an effort to deal with the crisis in the Coast Guard, on February 1, 1924, President Coolidge forwarded to Congress estimates of appropriations to a total of $13,853,989. The largest portion of this, $12,194,900, was to be dedicated to the reconditioning and equipping of twenty destroyers and two minesweepers in the United States Navy reserve fleet, and for the construction and equipment of 223 cabin cruiser–type motorboats as well as one hundred smaller motorboats. The twenty destroyers transferred from the Navy to the Coast Guard (joined by five more in 1926) had seen hard use during World War I and were part of a large fleet declared surplus following the signing of the armistice in November 1918. They'd been laid up for five years and, as a consequence, were in various states of disrepair, which required a tremendous amount of work to recondition them for sea again. In the end though, all these destroyers ended up serving in the US Coast Guard on the east coast and none were delegated to the west coast.

For the second line of defence, smaller vessels were to be used for patrolling closer to shore: in harbours, bays, inlets or large rivers like the Columbia. Paying close attention to the tactics and methods rum runners relied on, two classes of vessels were required. The first was a patrol boat of sufficient size and seaworthiness to monitor and picket offshore as well as throughout coastal waters. The chosen design was used for a fleet of 75-foot patrol boats, which came to be known as "six-bitters." They were constructed of wood, 75 feet in length with their hulls of sturdy construction and flush decked, with pilot houses and two trunk cabins. The machinery consisted of two six-cylinder Sterling gasoline engines capable of delivering two hundred horsepower each and able to drive the new cutters at fifteen to sixteen knots. But more importantly, they were well armed with a one-pounder mounted forward, a Lewis machine gun aft as well as a complement of small arms. Two hundred and three six-bitters were built in total, the first commissioned in October 1924 and the last in July

The 125-foot patrol vessel *Montgomery* was one of the dollar-and-a-quarter boats, as they were called by the Coast Guard. This photo was taken from the deck of the mother ship *Malahat*. Hugh Garling collection.

1925. In December 1924, it was announced that thirty of these new vessels that were being built in Seattle and Puget Sound Naval Station at Bremerton were to be assigned to patrol work throughout the Pacific Northwest. Twenty were for Puget Sound waters while seven were to operate in Oregon waters.

Also constructed for the Coast Guard were the "dollar" and "dollar and quarter" 100-foot and 125-foot boats. Capable of thirteen knots, these cutters were found to be excellent sea boats and were mounted with a three-inch twenty-five-calibre quick-firing gun and Lewis machine guns. Still, it was the six-bitters and 103 smaller picket boats that were built at the same time that became the backbone of the Coast Guard's inshore fleet.

The picket boats were of two designs: thirty were single cabin, open cockpit 35-footers while seventy-three were double cabin

36-footers, both types capable of twenty-four knots. Another source of reliable craft was the rum runners themselves. If captured and seized, they were sometimes transferred to the Coast Guard, converted and used to good advantage, as many were purpose-built to be fast and effective in protected waters, and a few were even seaworthy enough for offshore duty too. To man this rapidly expanding armada, the Coast Guard force of 5,982 personnel jumped to ten thousand between 1924 and 1926.

Congress had also authorized the additional personnel desperately needed by the force. As a result, it was only a short time after Commandant Billard's declaration that the service would put the government money to good use and the investment began to pay off. In less than three months, the Coast Guard could claim that seizures of liquor, along with the vessels carrying it, had jumped to more than ten million dollars in value which was nearly the same amount of money dedicated to the rum war at sea. But when Billard provided a year-end review of the Coast Guard activities for 1924 and pointed to the Coast Guard's overall success, he was still required to qualify that claim.

He admitted that while Rum Row no longer lingered off the coast at set locations, ships of various flags were still able to operate and remained very active in the trade by altering their modus operandi. Now, instead of being strung out in fixed locations along the American coast, the mother ships constantly moved about in order to make themselves harder to track. This development brought with it more frustration for the service. As it was, their most effective tactic was to stand by and keep suspected liquor-carrying vessels under close surveillance. This was a particular annoyance for the smaller local American operators since it placed them in a difficult situation if they tried to come alongside a mother ship to take on a load and then make a quick dash for the beach. (This was soon recognized throughout the rum fleet as "cutterizing" and if a particular Canadian shore boat was unfortunate enough to be cutterized for days or weeks with her skipper consequently becoming overly frustrated and frantic, he was said to have had a bad case of "cutter-itis.") If the Coast Guard was successful in making a capture, it often happened that after the case went to court, the legal tangle involved in trying to pin a charge on a rum runner and then attempting to make it stick, made many a lawyer wealthy, while those charged often walked away unscathed.

Through the late 1920s the bolstering of the US Coast Guard with more ships, men and equipment (as it was, by 1928, there were twenty-two Coast Guard vessels patrolling Puget Sound alone) greatly improved its operational effectiveness and there was a significant increase in its reliance on aggressive tactics. Fraser Miles concurred that what with the dramatic seizures of the *Quadra*, *Pescawha* and *Coal Harbour* through 1924 and 1925, along with the loss of *Speedway* to fire, the number of offshore rum vessels delivering liquid refreshments over to Yank firewater boats was so reduced that the Coast Guard cutters had little trouble limiting the activities of the few remaining ships.

Chapter Eight

THE CUSTOMS SCANDAL OF 1926

Throughout the early 1920s, Canada's federal government generally remained aloof from the American government's frustration with trying to enforce the Volstead Act. As it was, Canadian distilleries and breweries were on a roll, pumping out millions of gallons of liquor and beer into a parched US market. While the United States' neighbour to the north provided both the operating base and the raw materials for what was a multi-billion-dollar criminal industry, for a long time Canadians generally, and their government particularly, remained aloof and turned a fairly indifferent eye to what was going on. Some estimates had it that around five million gallons of liquor were smuggled into the US in 1924 and, as a result, the administration of the Canadian Customs and Excise Department soon became the *bête noire* of William Lyon Mackenzie King's Liberal government.

Early in 1925, when the Liberal minority government was appraised of the inefficiency and corruption that existed throughout the department, it launched an informal investigation to look into the matter. When the King government appeared reluctant to act on its findings, H. H. "Harry" Stevens, the Conservative member of Parliament for Vancouver Centre, finally stood up in the House of Commons on February 2, 1926, and, in a four-hour-long tirade, detailed the inside information that had been leaked to him. Stevens charged that the government was well aware of flagrant violations of customs regulations, that a number of officials in the

Customs and Excise Department were involved in wrongdoing and that corruption was endemic throughout the department. He even accused Jacques Bureau, its minister, of destroying nine filing cabinets crammed full of incriminating documents.

In response to his attack, the House was forced to create a special nine-member committee to investigate the administration of the department and the alleged losses to the public treasury because of inefficiency or corruption on the part of officers of the department. After 115 sittings and testimony from 224 witnesses, the final report of the parliamentary committee was completed and ready to present before the House on June 18, 1926. It confirmed that there was indeed widespread inefficiency and laxity in the department (some senior employees were found to be so delinquent in their duties that the report recommended the dismissal of nine of them) and that stolen automobiles were smuggled into Canada while liquor was being smuggled into the United States. In his book, *Iced: The Story of Organized Crime in Canada*, author Stephen Schneider said that the committee even went so far as to implicate the minister of customs and excise himself. The evidence revealed that the department, along with its deputy minister, had a large quantity of seized liquor transported from Canada customs warehouses in Montreal to their homes in Ottawa. As C. P. Howell, the customs officer in charge of the Roosville, BC, border crossing recalled, "It was not the rum-runners who corrupted us, but the Canadian politicians, federal, provincial and municipal."

Realizing full well that his government would lose in a vote in the House once the damning disclosures into the depth and scale of the customs scandal were presented, King resigned as Canada's prime minister on June 28 and asked Governor General Lord Byng for dissolution of Parliament and for an election to be called. But Byng refused, and the next day he asked Conservative leader Arthur Meighen to form a government, thereby stimulating a major constitutional crisis. Byng overstepped by refusing King's request for an election call and instead handing power over to the Opposition. Consequently, the new Conservative government fell three days later and a general election was called. Still, the Meighen administration ensured the passage of a censure resolution during its brief time in power. On June 29 it received unanimous support in the House to add a short statement to the report of the parliamentary committee: "Since the parliamentary inquiry indicates that the smuggling evils are so extensive and their ramifications so far-reaching that only a portion of the illegal practices have been brought to light, your committee recommends the appointment of a judicial commission with full powers to continue and complete investigating the administration of the Department of Customs and Excise and to prosecute all offenders."

On July 20, 1926, the Royal Commission on the Department of Customs and Excise was established by an Order-in-Council. Although bootlegging (delivering up illegal liquor by land) was to be its primary focus, it was also mandated to look into the liquor smuggling controversy. Meanwhile the election campaign was underway, with the overly confident Tories expecting the election to be fought over the customs scandal. King, on the other hand, conducted the Liberals' campaign on what they considered was the more serious constitutional crisis. (Lord Byng, as Dave McIntosh described him in his book *The Collectors: A History of Canadian Customs and Excise* "was a high-born Briton deciding on his own, like some old-style colonial governor, when and how elections should be called.") On election day, September 14, 1926, the "King-Byng affair" proved to be of more consequence to the electorate than the customs scandal and the King Liberals, although they won fewer seats than the Conservative Party, were able to govern with the support of the Progressive Party.

The Royal Commission began travelling the country in November that year and heard more than fifteen million words of testimony by the time they were finished in September 1927.

Meanwhile, a month earlier, in a letter dated October 9, 1926, that the customs department delivered to the RCMP's director of criminal investigations, an undercover agent provided detailed background about how West Coast exporters were redirecting liquor ostensibly consigned to foreign destinations: "Large shipments of liquor in transit ('in transitu') between Europe and Central American ports are landed at Vancouver and held in Sufferance sheds until convenient to trans-ship to certain boats engaged in rum running ... Among water front employees, such as Stevedores, Checkers, and Police, with whom I conversed, it is the opinion that a large quantity of the trans-shipped liquor never reaches the foreign Ports to which it is consigned, but is landed in Canadian waters, and ultimately returns to Vancouver for distribution among local bootleggers by such firms as Consolidated Exporters Corporation Ltd, Manitoba Refineries Ltd., Kennedy Silk Hat Co., Henry Reifel, R. T. Morgan, and other prominent men connected with distilleries and breweries in B.C. are freely spoken of as being the leading men in these activities."

In *My Dad, The Rum Runner*, the biography of his father, Captain Stuart Stone, author Jim Stone said that when bonded liquor supposedly destined for Mexico or Central America was smuggled back into Canada after being landed and unloaded tax and duty free, it could be subsequently sold at a discount to corrupt local buyers. Overall though, the Royal Commission's third interim report revealed that "enormous quantities" of liquor were shipped in transit without any duty paid: "The traffic has been carried on by fictitious consignees, clearances on false declarations as to destination, false return clearances and false landing certificates." And on the recommendation of the commission, the government passed legislation declaring that no clearance document was to be issued unless a customs collector received a bond for double the amount of customs duties payable.

Philip Metcalfe explained how in transitu worked in his biography of Roy Olmstead: "Merchandise shipped from outside Canada to parties outside Canada technically never entered Canada and thus never came within the jurisdiction of Canadian customs. On paper all Consolidated had to do was to pretend that its European liquor [buying] agents in London and Glasgow were independent exporters shipping liquor through Victoria and Vancouver to Mexico." The illicit liquor trade rose to an art, since "on paper the arrangements were made at both ends and Consolidated, like a magician's knot in a string, all but disappeared ... In transitu shipments not only enjoyed the $20 savings per case but required no bond guarantee." Still, "the bureaucratic edifice of Canadian Customs swelled to the size of a paper palace" since the documentation "specifying carriers on mixed loads for each step of the long journey from Scotland to Mexico increased the mysteries of the paperwork tenfold."

It also provided another loophole that Canadian distillers were able to exploit. Federal law specified that all Canadian liquor, which, of course, wouldn't end up in a bonded warehouse, was required to age for two years at the distillery unless exported. In order to circumvent this requirement, Canadian liquor companies, like those affiliated with Consolidated Exporters, would ship their freshly distilled product to connections in London, Glasgow and Antwerp and then have it shipped back to Canada. This practice, which was used to avoid restrictive legislation, was known as double-stamping. And as Metcalfe pointed out, "The additional transportation costs were minimal, and the quality of the liquor apparently benefited greatly from the rolling sea voyage."

In his tale, "Rum Row: Western," which appeared in the May 1932 issue of *American Mercury* magazine, Robert Dean Frisbie provided a supposedly fictional account of how in transitu was actually carried out. He suggested we consider a distillery located in Canada and only fifty miles from the border where, on the US side, a farmer is eager to get his hands on a bottle of Old Crow made in that distillery. So how does it actually get delivered to this farmer? "The distillery ships it to Halifax, where there are agents to pay, and warehouses and handling expenses and duties. From there it is shipped to a European port and is unloaded and put in bond. All this is paid for, and the European authorities probably get some graft out of it, too. Then there is another vessel, and more handling and wharfage and export duties, and the bottle of Old Crow goes through the Panama Canal to an island in the South Pacific Ocean— thousands of miles since it left the distillery! It is unloaded; more stevedores, bond and tax dues, storage and agents at the island branch; and then, by and by, more stevedores again, and the Old Crow is loaded on a schooner and taken nearly four thousand miles to the coast of Mexico.

"There it is trans-shipped aboard the five-master—-and that is a devil of an expense—with heavy breakage and demurrage, and a big expensive crew on the five-master to be paid double wages and messed at a dollar and a half a day. And the Coast Guard indirectly increases the expense of the whiskey, for buying the Old Crow makes Prohibition agents ashore and afloat necessary, and that increases taxes. Presently it goes aboard a sub-chaser—more expense—and is transhipped again, this time to a speedboat off the coast of California ... Well, the speedboat takes it ashore to a truck, and the little bottle of Old Crow finally rests for a while in a big wholesale warehouse of one of the shore operators. But not for long. Presently the farmer up in Washington wants his bottle of Old Crow, so he orders it through his bootlegger, and the next morning finds it under his doormat. Now, I wonder if it ever occurs to him that the bottle he is drinking has gone more than half way round the world before reaching him—while it was made in a distillery a few miles from his farm!"

Later, in one of its 1928 interim reports, the Royal Commission explained how the Reifel family managed Joseph Kennedy Limited, a holding company that operated a number of bonded export houses. This holding company shared office space with the Kennedy Silk Hat Cocktail Company in the Gray Block in downtown Vancouver,

The Gray Block, 1206 Homer Street, downtown Vancouver, BC, circa 1928. City of Vancouver Archives, AM54-S4-: Bu N288, photo by W.J. Moore.

with Reifel's sons George and Harry serving as president and general manager. (This particular Kennedy was not Joseph P. Kennedy, father of future US president John F. Kennedy. Vancouver's Joseph Kennedy was American, but from Missouri, not Massachusetts. Rumours that still circulate have it that the Massachusetts Kennedy may also have been involved with smuggling during US Prohibition.) The report noted that the sole business of Joseph Kennedy Limited was "the export of liquor to the United States" and that there were "a great many irregularities, some of serious nature, in connection with this company."

When he gave his testimony before the Royal Commission, Henry Reifel candidly admitted to providing nearly one hundred thousand dollars to British Columbia's provincial political campaigns in 1925 and 1926. He informed the commission that, in return, he didn't receive any "promises," although the payments were entered in the company's financial ledger "as assurances and protection." Still, Reifel, apparently fully cognizant that his transgressions were now in the public domain, recommended that the commission consider a law prohibiting campaign contributions since his experience was "you never get any return on the money."

The issue of forged landing certificates had come up in one of the Royal Commission's interim reports. The commission noted that mother ships departing from Vancouver and Victoria often gave false destinations outside the US, but actually unloaded off the US or Mexican Rum Row. They would bribe corrupt officials at their stated destination to forge papers showing that the cargo was actually unloaded there, then somehow get the forged paperwork back out to the mother ship before it returned to Canada.

As Dave McIntosh described in his history of Canada's customs service, "The government was not swift to crack down on forged landing certificates, possibly because taxes on liquor provided, as they do today, considerable revenue (in 1929, liquor taxes brought in $60 million) twice as much as the personal income tax." US Prohibition officials were quick to realize that their efforts to curtail a flood tide of rum running were to be constantly frustrated and blamed it all on the harsh reality that just across the border, the Canadian government looked the other way in order to enrich its coffers from the production and export of liquor. It was only after the final report of the Royal Commission was released that Canada's federal government passed legislation to make it appear as though it was finally making a more serious attempt to put an end to liquor smuggling. On June 1, 1930, legislation was passed that made it illegal to export liquor to any nation that prohibited the importation of liquor.

Regardless, well-established entities in the trade such as Consolidated Exporters had already made prior arrangements to circumvent the law and were using Papeete, Tahiti, as a transshipment port rather than Vancouver and Victoria. As Revenue Minister

W. D. Euler was to inform the Canada's House of Commons on May 21, 1929, "It is impossible to have wet and dry countries adjacent to each other without a flow from the wet to the dry." Thanks to its geography, Canada especially benefited from this arrangement. With a four-thousand-mile border adjoining its neighbour to the south and oceans at either end, it indeed proved a smuggler's paradise.

SEIZURE OF THE SCHOONER
CHRIS MOLLER

Back in its first interim report delivered on December 3, 1926, the Royal Commission on the Department of Customs and Excise provided "a typical illustration of what is and what has been a practice at the ports of Vancouver and Victoria in connection with the so-called shipments of liquor." The example it chose to highlight the prevailing type of fraud being perpetrated was that of the wood five-masted schooner *Chris Moller*. It may be that the commissioners chose this particular mother ship since she had just loaded a substantial cargo of liquor in Vancouver at the time the interim report was wrapping up.

According to the December 28 *Daily Colonist* article that stated *Moller*'s cargo was to be seized, "the destination of the cargo, according to declarations made on the ship's papers, was San Blas, Mexico, but witnesses before the customs commission stated that the liquor was destined for consumption in Los Angeles and California."

While the newspaper accounts of the day referred to the vessel as *Chris Moeller*, the Lloyd's Register of 1924–25 lists her as *Chris Moller*. Like the *Malahat*, she was an auxiliary lumber schooner built during the World War I shipbuilding boom, but unlike the *Malahat*, she was not built in a Canadian yard. At 266 feet in length and built for a Norwegian owner, the *Chris Moller* was launched from the Olympia Shipbuilding Company shipyard in Olympia, Washington, in July 1917 as the *Wergeland*.

In the fall of 1917, setting out on her maiden voyage carrying lumber to Norway via Portland, Oregon, *Wergeland* was battered by heavy seas in a storm off Tatoosh, at the southern entrance to Juan de Fuca Strait, and lost two masts and her deck load while sustaining serious damage to her hull. She was towed back to Port Blakely, Washington, for repairs. Sometime soon after, the schooner was sold off.

According to her entry in the 1921–22 Lloyd's Register, *Wergeland* was sold to "Moller & Co (Shanghai) Ltd," and her flag was British. In

February 1924, Archie MacGillis purchased the schooner and had her reconditioned by the Shanghai Dock & Engineering Company. In June of 1924 he registered *Chris Moller* in China's big port city under the name of one of his liquor export shipping companies, Iron Bark Exchange. On July 27 she cleared Shanghai for Vancouver.

In 1966, M. P. Olsen wrote to Charles DeFieux, marine editor of the *Vancouver Sun*, and passed on the tale of how *Chris Moller* "loaded a cargo of liquor (and Chinamen) in Shanghai" and subsequently got caught in a typhoon and returned to port with a two-foot permanent twist in her hull. After she was dry-docked and repaired, MacGillis went over and put her back in commission. The paper reported that the new boat, with Australian Arthur G. Lilly of Vancouver listed as master, was to carry a liquor cargo to Topolobampo, Mexico, and then return to Vancouver to load lumber for Asia, including "cedar logs for Chinese coffins." Rumours suggested that the owners were actually planning a wholesale importation of opium and cocaine, as well as whisky, but these were dismissed by many as bogus tales being spread by their rivals in the liquor export trade.

Chris Moller maintained a low profile in the news over the next two years, especially with the capture of other mother ships throughout 1924 and 1925 grabbing the headlines. A short article with a photo of the schooner finally appeared in the June 5, 1926, edition of Vancouver's *Daily Province* under the headline: "Will be the Largest Halibut Craft Working on the Fishing Banks of the North Pacific." The paper reported that the *Chris Moller* was "being fitted out at the BC Marine yards with four fish-freezing units of seven and one-half tons each, and a cold room with a capacity of 1,000 tons. She will carry thirty fishing dories and a powerful launch tender ... All that comes to her nets and hooks will be fish." MacGillis always had a good story prepared beforehand if he was to be interviewed by the press, so whether *Moller* was ever fitted out for fish packing is doubtful. (MacGillis always remained cagey regarding his involvement with the liquor export trade. The 1928 edition of *Wrigley's British Columbia Directory* indicates that the managing

Chris Moller was seized after American Prohibition authorities informed Canadian officials that the small town of San Blas, Mexico, where the schooner was destined with 55,000 cases of Scotch whisky aboard, had a population just large enough to enjoy fifty-five bottles. Rick James collection.

director of Iron Bark Exchange Limited was listed as "A. MacGibles.") Iron Bark Exchange kept *Chris Moller*'s Shanghai registry right up until her demise.

Sometime in the final months of 1926, *Chris Moller* cleared the port of Vancouver with 17,779 cases of liquor aboard imported from Great Britain by Manitoba Refineries, a BC liquor export company. It then called in Victoria to load an additional 3,700 cases, but it was here, alongside the Ogden Point piers in Victoria, that customs officials refused clearance pending an investigation.

Customs and Excise officials had become suspicious, since declarations made on the ship's papers stated that the cargo's destination was San Blas, Mexico, which happened to have no port facilities. As it was, the liquor was imported from Great Britain to Vancouver by way of the Panama Canal, past that small village along the Mexican coast on its way north, though it never stopped there.

The customs report concluded that "the alleged consignee (a 'Mr. Rodriquez') is fictitious and that it is not intended that the liquors should be delivered [by the *Chris Moller*] at San Blas, the port of destination, but rather that the same should be made available elsewhere to rumrunners or bootleggers for consumption in the Western states." Witnesses in Victoria said that the story out on the street was that the liquor was most likely destined for San Francisco and Los Angeles, but had been delivered to Vancouver first, where it went into a bonded warehouse to avoid taxes.

On December 27, 1926, a wire was sent to Collector Fred W. Davey from the acting deputy minister of Customs and Excise in Ottawa, ordering the seizure of the cargo and the subsequent release of the ship. The reason given, according to the December 28 *Daily Colonist*, was that they had nothing against the ship other than the false declarations that were made in regard to her cargo.

Early in January 1927, *The Daily Colonist* reported that before the liquor was completely discharged at the Ogden Point docks, thirty-six cases of the cargo "vanished into thin air, but customs officials would, no doubt, like to get hold of the 'Houdini,' who managed to cast a spell over the contents of the other ten cases, which have been opened and only broken and empty bottles left." The next day the paper reported that Chief of Police John Fry and Deputy Chief Harry O'Leary managed to arrest *Moller*'s Chief Officer Warrington along with "an Italian named Angelo Sposito and Mrs. Marion Sposito for receiving liquor which they knew to have been stolen."

The *Chris Moller* was released on an eight-hundred-dollar deposit and on June 4, 1927, departed Vancouver on her next and final deep-water voyage with Arthur G. Lilly as her master. She had a full load of lumber loaded at the H. R. MacMillan mill at Ladysmith on Vancouver Island and was destined for Newfoundland. She arrived in St. John's on August 23. Once she discharged there, she left in ballast for Antwerp, Belgium, where she was to load liquor. But as M. P. Olsen told Charles DeFieux of the *Sun*, after arriving in Antwerp on November 1, 1927, and failing to arrange for such a cargo, *Chris Moller* ran up a heavy shipyard bill, which resulted in the yard seizing the ship and the Canadian government having to foot the bill to return her master, Arthur Lilly, and crew to Canada. As for the schooner herself? She was broken up in Belgium shortly thereafter.

Chapter Nine

"HAIL, ALE, THE GANG'S ALL HERE"

CAPTAIN STUART S. STONE AND CAPTURE OF THE *FEDERALSHIP* (1927)

Hail, ale, the gang's all here
Going down to run a cargo,
Way down to Ensenada.
Uncle Sam will have his cheer,
And Volstead we must beat, somehow ...

Over there, over there, bully gee,
There's a cutter over there.
Look out, she's coming,
Knows we're rum-running,
She's sure got speed, I do declare,
Speed her up Sandy Mac,
Make your Fairy Morsey Engine fairly crack,
Now it's over, let's thank Jehovah.
I've dropped a case, stopped the chase,
Jackeroo! She's going back ...

Never say that Frisco is a one-horse town,
Now let it trickle right down your throat,
Every time you swallow you get Volstead's goat,
For I am the skipper of a small speed boat,
Shooting [hootch] *in every day.*

— "THE RUM RUNNER" BY "SIDNEY V. ELVY," FOUND IN THE CABIN
OF CAPTAIN STUART STONE BY PROHIBITION OFFICIALS

Victoria residents opened their *Daily Times* on March 4, 1927, to learn that a "Vivid Story of Capture of Rum Runner" was inside. News from San Francisco reported that a US Coast Guard patrol vessel, a "Bull Dog of the Sea ... the little craft *Algonquin*, spic and span speedster of the patrol, working out of Astoria ... had just written a new and vivid history on the Pacific in the last few days."

After returning to the small Oregon port after a futile attempt to save the lumber schooner *Mary E. Moore*, which went down before they could reach her, *Algonquin*'s commander, Lieutenant William S. Shannon, was ordered to put to sea again without delay and search for the rum runner *Federalship*. There was so little time in the quick turnaround that according to the chief engineer, Lieutenant B. C. Wilcox, "within two days cigarettes were selling for 25 cents a piece on board, and butts were at a premium."

The *Federalship* was an iron screw steamer launched as *Campine* by G. Howaldt of Kiel, Germany, in 1884. She measured 222 feet, 8 inches in length. A compound steam engine provided eighty-five horsepower. While still under German registry as the steamer *Arnold*, she was captured by the British during World War I and saw little activity until awarded to Belgium in postwar reparations and renamed *Gertrude*.

Hugh Garling said that *Gertrude* left Scotland for Rum Row loaded with a cargo of liquor and expecting to realize good sales off the American coast, but after a year failed to make a sale since she was consistently undersold by the well-established organizations working out of British Columbia and Washington state. Captain Charles Hudson said the steamer arrived with around twenty thousand cases of Scotches and other liquors and remained off the Farallones for six to seven months with no contact from shore. *The Vancouver Daily Province* later noted that when she arrived from Europe in November 1925, *Gertrude* experienced a shortage of food supplies while lying off San Francisco for several months waiting to dispose of her cargo. And "on reaching Vancouver port her sailors complained of sustaining life for weeks on the flesh of seagulls."

Consolidated Exporters finally bought the forty-three-year-old steamer, along with her cargo, while she was still sitting off the American coast, and had all the liquor transferred over to the *Malahat*, which was also out on Rum Row at the time. After arriving in Vancouver in ballast, she was overhauled, renamed *Federalship* and registered as owned by the Federal Shipping Company Limited, of Vancouver, a Consolidated Exporters subsidiary, which put her to work as a mother ship. Charles Hudson, who went on to become marine superintendent manager, or "shore captain" of Consolidated Exporters not long after the *Coal Harbour* trial was settled, told Ron Burton in his 1960s CBC interview that Consolidated Exporters "didn't give a damn for the federals and they called her out of a sense of bravado and rechristened her *Federalship*."

By early 1927, *Federalship*, with Stuart Stone as master, was under charter to Archie MacGillis's Canadian-Mexican Shipping Company, which was operating Consolidated's ships-at-sea operations. *Federalship* was sailing under a foreign flag at the time, since she been registered at the Panamanian consulate in Vancouver in April of 1926, which was finally confirmed by the Central American nation four months later, on August 25. (As Jim Stone noted in the biography of his father, there was always a deliberate blurring of ownership—a cover-up in the books—just in case anything untoward should happen.)

Jim Stone described *Federalship*'s captain and senior officers as some of the best seamen on the coast. Hugh Garling, who crewed under Captain Stone a few years later in *Malahat*, described him as "a short, heavy set, energetic man, but if short in stature, he was long in nerve, as he was soon to demonstrate." Stuart Stone was born in Halls Bridge, Ontario, in 1888 to a local teacher, William John Stone, and his wife, Christina Irwin, and moved out to the West Coast with the family as a young boy in 1891. William, a Methodist who was not satisfied with the teaching profession, started studying theology by correspondence through the University of Toronto, and his church sent him out to the Nass River area in northern BC to practise his missionary skills. Upon receiving his ordination three years later, he was posted to Clo-oose, an isolated village on the west coast of Vancouver Island, to establish a Methodist mission among the local Dididaht First Nation. In 1904, the family moved to the Tofino area, where William founded the Methodist church and hospital out on Stockham Island, while Christina set up a prosperous dairy. In 1915, they all settled in Port Alberni.

By 1920, Stuart Stone was out working on boats with his father, William Stone, and Stuart's brother Chester ("Chet"), first on the 37-foot motor vessel *Tofino*, and then with the former North Arm Fraser River ferry, the 75-foot, 6-inch *Roche Point*, transporting passengers, mail and general freight out Barkley Sound to Bamfield, Ucluelet and waypoints. Seeing an opportunity to make better money, the Stone brothers were quick to jump into rum running as soon as it got rolling in the early 1920s. According to long-time Tofino resident and local historian Ken Gibson, Stone and his brother Chet started out smuggling bottles of whisky in the *Roche Point* from the west coast of the island by hiding them under cargos of rotten chum salmon they were packing into Seattle harbour for use as mink feed. One night the gentleman they were to supposedly deliver their "fish" to lit out on them, leaving them cashless and unable to pay harbour duties. Slipping by the guard posted to the boat to prevent it from leaving port, Stuart and Chet snuck aboard in the middle of the night and took off back to Vancouver Island before the Coast Guard, who they suspected were probably well aware of what they were up to, arrived on the scene. Sometime in late 1922 or early 1923, after he signed on as captain of the 87-foot, 7-inch halibut boat *Kiltuish*, owned by Consolidated Exporters, Stone had to admit to having officially joined the brotherhood. He moved into rum running big time when he took charge of the 190-foot motor ship *Principio*, registered to Archie MacGillis's Canadian-Mexican Shipping Company around the time sister mother ships *Quadra* and *Coal Harbour* were both captured. (His son Jim Stone said that while this was all going down, his dad maintained a low profile.)

Captain Stone's first officer and supercargo on *Federalship*, James F. "Jim" Donohue, had gone to sea as a boy and been around the globe a few times on windjammers and steamships before joining the rum fraternity. Donohue, who was originally from St. John's, Newfoundland, was already a seasoned rum running mariner, having served as second officer under Captain George Murray in *Malahat* in a voyage that lasted from November 1924 through to August 1925. Also, *Malahat*'s official Agreement or Articles and Account of Crew for a voyage, "Vancouver to Mexico and Return," which the thirty-nine-year-old Donohue signed, revealed that the last ship he'd served in was another five-masted auxiliary schooner, the American-built *Chris Moller*. Stuart Stone said that he first met Donohue in Tahiti, "where he'd gone native a la Gauguin," after arriving in there on a dry run with *Federalship*. He was able to persuade Donohue to join the *Federalship* for some more adventure. As Jim Stone described Donohue, he was rather taciturn but he knew how to handle men. He and Stuart soon became fast friends, with Donohue even going on to marry Stuart's sister, Hazel. Hugh Garling, who crewed with Stone and Donohue aboard *Malahat*, concurred: "[Donohue] was fine a seaman as you would find. With a long scar the length of his left cheek, he had a piratical look, adding some mystery and a little reverence. The fo'c'sle stories among the seamen suggested a bar room brawl somewhere in the South Seas. Donohue … could wither you with a scathing glance alone."

Among the other crew were Second Officer Alexander Kennedy, and Stuart's brother, Chet Stone, serving as chief engineer. And as for the crew? Several of the deckhands, while only teenagers, soon learned that they were in good hands and could count on their

Captain Stuart Stone (left), pictured here with First Officer Jim Donohue (right), was already a deep-sea mariner freighting out of Port Alberni on Vancouver Island's west coast when he decided to expand his horizons by moving into rum running. After *Federalship*, Captain Stone went on to take command of her mother ship, *Malahat*, in 1929. Fred Sailes collection, courtesy of Valerie Allen.

officers, all well-seasoned mariners. As Hugh Garling described, "Both Captain Stone and his Chief Officer were loath to indulge in aimless pleasantries with any of the crew always maintaining an aloofness which reinforced their command. With a crew of 18 men in a ship carrying a million-dollar cargo of 12,500 cases of whisky, if discipline and good order were not maintained they would be sitting on a powder keg."

After coming alongside Ballantyne Pier sometime in mid-February, *Federalship* loaded 12,500 cases of choice in-transitu whisky and imported wines shipped out from Glasgow (worth an estimated one million dollars). With Captain Stone in command, the mother ship cleared Vancouver harbour on February 22, 1927. According to her manifest, for all intents and purposes, ship and cargo were destined for Buenaventura, Colombia.

Once the Coast Guard ship *Algonquin* sighted the steamer wallowing in the heavy swells seventy-five miles off the mouth of the Columbia River, the little craft clung to its quarry day after day and followed the rum runner to a point three hundred miles off San Francisco. Several times the steamer doused its lights and attempted to escape under cover. In a March 7 article, the *Daily Times* continued with its story of the "Vivid Capture": "It would speed up, then zig zag on its course, as though trying to throw off a war-time submarine, but to no avail. *Algonquin* kept right in its wash." Lieutenant William S. Shannon, commander of the cutter, was finally able to run alongside the old steamer, where he demanded that Captain Stone explain to him exactly what they were up to. "Oh, just cruising around" was the reply. It only took a few minutes for the Panamanian flag to be run up on *Federalship* and the order rung down to the engine room to put on more speed.

The cutter *Cahokia* joined the *Algonquin* on the scene on February 28 after being dispatched from San Francisco with a direct order to either sink or seize the mother ship and bring her into port. (The commander of *Cahokia* just happened to be Chief Boatswain Sigvard B. Johnson who had managed to capture mother ship *Coal Harbour* three years earlier.) As the *Daily Times* reported, a defiant Captain Stone refused to heave to and shouted across the water, "I'll not stop. I'm on the high seas and you have no right to stop me!"

By this time nightfall was setting in, so both Coast Guard vessels backed off to wait until morning. With the arrival of daylight on March 1, *Cahokia* once again ordered the steamer to heave to and when Captain Stone once again refused to comply, the cutter fired several blank shells from its one-pounder across *Federalship*'s bow and followed up with a live round while sister ship *Algonquin* raised the international signal flag ordering a vessel to stop instantly.

When *Federalship* kept on going, *Algonquin*, which had been steaming some distance astern and was armed with heavier guns, moved up and trained her six-pounders loaded with solid shot on the rum ship. Captain Stone stood his ground, shouting across the water through a megaphone, "I'll not stop. How do I know but that you are not a lot of bloody pirates!" as the *San Francisco Chronicle* reported in its lengthy story on the dramatic capture.

In response, Ensign Frank K. Johnson, the cutter's division officer aft, was ordered to "let go." "Stop in the name of the United
States government! For the last time, I order you to stop!" But
Captain Stone refused and yelled back across the water, "in the
nyme of jehovah, h'im hon the 'igh seas and h'll not stop!" The
paper continued, quoting Lieutenant William S. Shannon, master
of *Algonquin*: "one shot went across the bow of the speeding quarry.
Still she kept going. The crack crew of the Bering Sea fleet, of which
the *Algonquin* is flag ship, then fired two shots from the six-pounders." In an attempt to shoot away the rudder and steering gear, the
second shot hit the starboard rail and blew off the forward hatch
cover, throwing splinters all over the stern only some twenty feet
from where Stone was standing on the bridge. When the smoke
cleared, Captain Stone, fearing that his old ship with its thin steel
plates was about to be sent to the bottom, raised his hands high in
the air after ringing down "stop engines," letting *Algonquin* know
it was all over.

At this time *Federalship* was lying 270 miles off the territorial waters of the United States. In its reporting on the seizure in
its March 2, 1927, edition, Vancouver's *Daily Province* stated that
it was believed that the Canadian and United States governments
had arrived at an agreement, whereby the Canadian government
consented to aid the United States in putting a stop to liquor smuggling by allowing the seizure of vessels working under Canadian
command and ownership on the high seas for alleged past offences
under the Prohibition Act. Still, "it thought that the *Federalship* was
taken on some old score. The legal definition of 'pirate' could not be
applied to the vessel in any way, shipping men say ... It was Uncle
Sam's first use of his asserted right to treat a known rum runner like
a known pirate ship by seizing the vessel anywhere that it may be
found, regardless of three-mile limit, twelve-mile limit or any other
restriction." It further noted that following this supposed agreement between the Canadian and American federal governments,

After Consolidated Exporters Corporation bought the old steamer *Gertrude* in 1926, they renamed her *Federalship* as a way of thumbing their nose at federal authorities. Here, an unnamed us Coast Guard cutter lies alongside. *Federalship*, March 9, 1927, AAE-1357, in Ships folder, Box PxS22, San Francisco History Center, San Francisco Public Library.

Consolidated Exporters announced that it intended to quit Canada and instead establish its base of operations in Tahiti.

With *Federalship* hove to and with Captain Stone and his super-cargo and first officer, Jim Donohue, refusing to cooperate with their boarders, *Cahokia*'s captain placed them under arrest and escorted them aboard his craft while the eighteen members of the mother ship's crew were herded below deck. A prize crew took control of the vessel and the long tow was soon underway into San Francisco Bay under convoy of four cutters (the two other cutters on the scene were *Shawnee* and *Smith*). Ironically, *Federalship*, with its unconventional name, was now under the jurisdiction of the federal government of the us. (A crewman later recounted how several of the Coast Guard prize crew got down into the mother ship's hold through the shattered hatch. There they helped themselves to some of the cargo and ended up thoroughly inebriated by the time the cutter docked in San Francisco.)

Following her seizure and arrival in San Francisco Bay on the towline of the cutter *Cahokia*, *Federalship* was anchored off Hunters Point in San Francisco Bay. Here, she was boarded by a party of federal authorities consisting of United States Attorney George J. Hatfield; Eugene Bennett, his chief deputy; William McBride, chief special agent of the customs service; and Alf Oftedal, head of the local intelligence unit of the Internal Revenue Bureau, along with seven of his operatives. Upon close inspection of the rum ship, they discovered that the cargo, on the whole, was still intact. All 12,500 cases were later removed to a US customs appraisers' warehouse while Captain Stone was to be arrested on a formal warrant on March 5, 1927. While the officials took the ship's papers and samples of her cargo, a small arsenal of Winchester rifles, revolvers and ammunition was left aboard but under guard. Still, much to his chagrin, Harold Faulkner, one of the attorneys working for the "Canadian syndicate," Consolidated Exporters (Attorney C. S. Arnold was representing the vessel's owner, Federal Shipping Company Limited) was not permitted to board while all this was going on. (Faulkner, as it happened, also represented the defendants in the *Quadra* and *Coal Harbour* trials.) "I was denied the right to see my clients, but I can not be denied the right to express an opinion. That opinion is that the coast guard exceeded its authority in seizing this ship 300 miles off shore" he told reporters.

When he was taken before a United States Commissioner, Captain Stone's bail bond was set at twenty thousand dollars on a charge of conspiracy to violate the customs law and he was remanded to jail on default of bail money. (A search of the mother ship revealed that the versatile skipper was also a wireless operator and kept a complete broadcasting and receiving station on his ship in perfect order.) First Officer James Donohue, along with

Second Mate Alexander Kennedy, were also held on bond for the same amount, as were all the rest of the crew, who were detained on board *Federalship*. The *San Francisco Chronicle* noted that after sixty-one charges were brought against her skipper along with "each and every man from millionaire liquor dealers to the lowliest truck driver and deckhand" the total bail on all counts amounted to $1,200,000. And after a *Chronicle* reporter went out to check up on *Federalship* anchored off Hunters Point, he wrote, "members of the crew wandered about and below decks in disgruntled groups, a conglomerate of whites, yellows and blacks of six nationalities although all claim to be British subjects." Three days later, the paper pointed out that all the defendants were "citizens of a foreign nation, viz, of the Dominion of Canada, and subjects of his Britannic majesty King George V."

However, photographs that covered the entire top half of page three of the March 3 *Chronicle* contradicted the reporter's opinion that the crew were a disgruntled bunch. One of the photos featured the crew lounging about on deck, apparently engaged in easygoing conversation. "The crew of the Federalship aren't worrying. They're laughing it off." The other photo was a rather charming portrait of "Captain Stone in a smiling close-up." A reporter with *The Vancouver Daily Province* elaborated by noting that "the crew of eighteen is composed mostly of young men, most of whom took their experience yesterday as a roistering, thrilling adventure. 'Get our pictures in the paper!' they yelled over the side of the ship." Uncle Sam's seizure of *Federalship* and the subsequent legal battle and shenanigans would continue to dominate big city news both in California and British Columbia over the next two months, much to the amusement of the readership. Editors gave the rather convoluted, but most entertaining international legal drama a lot of coverage and often dedicated columns on two or three pages to in-depth reports.

As the *Federalship* was registered in Panama, Captain Stone was quick to appeal for aid from the Panamanian government, which replied that they were still waiting to receive official information from their legation in Washington, DC. Stone argued that the Coast Guard boats that made the seizure had committed an act of war against Panama. The *Vancouver Daily Province* said that Stone told a reporter, "I never before heard of an American ship firing on a peaceful merchantman who had violated no law. Surely there must be some redress from such treatment." Captain Stone was more than obliging in sharing the details of all that transpired with newspaper reporters and told one from the *San Francisco Chronicle* that "when she [*Cahokia*] began following us and became so positively annoying I was compelled to ask what she was doing about. At one time she cut directly across our bow. It was so exasperating I asked her to please get out from under our feet and let us proceed. The blooming ship popped at us with a blank and next morning sent a round shot between our masts. I should not have stopped at all save that the other cutter [*Algonquin*] hammered us with a shot and I feared I should be swept off the deck. We were 300 miles out on the high seas and the seizure was an abominable act of war."

The pre-trial attack was already underway even before officers and crew of *Federalship* stepped ashore. Back on March 3, federal Prohibition director General Lincoln C. Andrews wired from Washington that "the seizure was made under the terms of a new Panama law which revokes the registry of any ship flying the flag of that republic, which resorts to smuggling and thereby leaves it a ship without a flag." Three days later, *The Daily Colonist* reported that a federal grand jury met in an extraordinary session and brought "the now-famous Federalship rum-runner case to an astounding climax today by indicting all of the members of the crew, three supposed directors of the Consolidated Exporters and a number of others on conspiracy to violate the Volstead Act and related laws and treaties."

The *San Francisco Chronicle* said that Consolidated was, in essence, "a gigantic liquor combine formed mainly for the purpose of getting contraband goods into the United States." The three directors indicted were R. Whitelaw, George Norgan and Albert L. McLennan, as well as Frederick Rae Anderson, an attorney representing Consolidated, and W. J. Murdock, the attorney representing the Federal Shipping Company. A lengthy statement of facts concerning the operation of *Federalship* off the US west coast was published by the State Department. Treasury agents reported, regardless of her flag, that there were no Panamanian citizens aboard the vessel at the time of her seizure, that her registered owner was a Canadian corporation that had flagged her in Panama and that Captain Stuart S. Stone was understood to be a Canadian who was under indictment in the United States in connection with previous smuggling operations. As US Attorney George J. Hatfield explained, "a court may proceed in a criminal case against a defendant even when the defendant has been brought before it by kidnapping him from a foreign country ... a pirate is outlawed by all countries and may be proceeded against wherever he may be found."

Besides being indicted on the charge related to the seizure of *Federalship*, Jim Stone said that his father was also brought ashore by US Marshal Frederick Esola and formally arrested on a prior charge of conspiracy to violate the Prohibition Act. It was in regard to a case involving the former mayor of Sausalito, J. Herbert Madden, and the town's local bootlegger, San Francisco tailor Joe Parente, as well as several others. Since there were few details on the case, Jim Stone suggested that it was no doubt related to how widespread and endemic it was to run liquor into San Francisco prior to 1927. Federal agents were attempting to crack down on municipal officials and customs officers who were enabling their own citizenry involved in the liquor trade, as well as Canadian rum runners who

Muse at Helm of Rum Ship

❖❖ ❖❖ ❖❖ ❖❖ ❖❖

Runners Stage Comic Opera

❖❖ ❖❖ ❖❖ ❖❖ ❖❖

Hail, Ale, Gang's All Here

Hail, ale, the gang's all here.
  Going to run a cargo,
  Way down to Ensenada,
Uncle Sam will have his cheer,
And Volstead we must beat, somehow.

So runs the opening lines of a long poem found in the room of Captain C. C. Stone of the rum-runner Federalship. The lines were printed on a strip of paper resembling a newspaper proof sheet. The "poem" is entitled "The Rum-Runner," and the author is Sidney V. Elvy. There is a cast of characters, from Captain Brew and his daughter, Benedictine, to Victor Cliquot, the cook. To the air of some of the popular songs of this and other days, each member of the cast or crew of the rum-runner sings his lines. Those by Benedictine to the tune of "Over There," are:

Over there, over there, bully gee.
There's a cutter over there.
Look out, she's coming.
Knows we're rum-running.
She's sure got speed, I do declare,
Speed her up Sandy Mac (engineer).
Make your Fairy Morsey Engine fairly crack.
Now it's over, let's thank Jehovah.
I've dropped a case, stopped the chase.
Jackerloo! She's going back.

The last four lines of the doggerel are:

Never say that Frisco is a one-horse town,
Now let it trickle right down your throat,
Every time you swallow you get Volstead's goat,
For I am the skipper of a swell speed boat,
Shooting bootch in every day.

There is another quotation—likewise with a Scotch tinge—heard around the Federal building. It is: "A man's a man for a' that."

And with the quotation is the gentle hint that the genial captain, himself, is the author of "Rum's Sweet Refrain."

A newspaper clipping from the *San Francisco Chronicle*, March 4, 1927. San Francisco Chronicle, March 4, 1927, courtesy of San Francisco Public Library microfilm files.

were landing and unloading with impunity in the smaller ports like Sausalito, if not right into San Francisco itself.

While three habeas corpus petitions asking for the immediate release of Captain Stone, First Mate Jim Donohue and crewmember Edward Solem went before Judge George M. Bourquin of the United States District Court, officers and crew of *Federalship* were all transported ashore aboard a Coast Guard cutter on March 7. As a reporter observed, all were in high spirits, singing "Hail, Ale, the Gang's All Here!" a rendition of a popular song they'd rewritten following their arrival in San Francisco. Once ashore and unable to come up with bail, officers and crew were locked up in the county jail, which became so overcrowded that extra cots had to be set up to accommodate them. They were to remain in the

"Bootleggers' College," as it came to be known during Prohibition, for almost two months. In his interview with Ron Burton, Captain Charles Hudson, who went on to manage Consolidated Exporters in the later years of Prohibition, said this was contrary to the practice maintained in the past, when they were usually all out on bail in a week. This time the attitude was "to hell with them, they all stayed in jail."

Hudson also heard that American agents were sent up to Vancouver, where they applied pressure to some of the former crew, especially those who had been fired for misdemeanour and might be more willing to open up and provide them information. The agents were particularly keen to learn anything about American boats that had been loading, to show *Federalship* had contact with shore.

Meanwhile, attorneys for Stone and his officers and crew challenged the court. Defence continued to argue that the US government acted outside its jurisdiction by arresting *Federalship*'s crew three hundred miles out at sea. It maintained that, "under our treaty with Great Britain it was only intended that this country should have the right to search and seizure of the ship. It was not permitted to permit prosecution of the crew [under the Volstead Act on a smuggling charge]." (It is thought that while the treaty was initially agreed to only between Britain and the US, it was considered an international law. Those were the days when Britain still "ruled the waves.") Government officials also disclosed that agents of their intelligence service managed to decipher the code used in radio messages to and from *Federalship* prior to the vessel's seizure and regarded the contents of the messages as important evidence. Still, Captain Stone remained adamant in his charge that the action constituted an act of war against Panama.

On March 14, the government of Panama finally got into the act. Ernesto de la Guardia, junior consul for the Republic of Panama, announced that his government's minister in Washington, DC, was demanding a full investigation. He noted that the Kellogg-Alfaro

Treaty between the United States and Panama, in effect since 1925, had been violated, and demanded to see all the official facts gathered by the US government. As it stood, the treaty only permitted search and seizure of any vessel within one hour's sailing of the US coast. Initially the defendants, of course, were very dismayed, since Panama was slow in stepping forward to defend their ship. But if the evidence proved that the treaty was violated, Panama's minister of foreign affairs was going to demand the US government release the vessel and crew as well as pay for all damages. Regardless, Consolidated Exporters went ahead filing its own protests and lodged them with the British foreign office, as well as its ambassador in Washington, DC, and its consul in San Francisco, and with the Canadian legation in Washington. (Even though most of the crew probably lived on Canada's west coast at the time, the majority of them were British subjects. As it stood as of January 1, 1915, all those born within His Majesty's dominions automatically became British subjects at birth.)

Two weeks later, Panama backtracked and requested the owners of *Federalship* forfeit the ship's Panama registry. Following the seizure of the *Federalship* on March 1, the *Chronicle* reported four days later that messages from Panama hinted that the vessel had already been under suspicion and warned once that her registry would be cancelled unless "certain operations ceased." Then, according to a news brief from San Francisco run in the April 2, 1927, *Daily Colonist*, federal officials in San Francisco "declared the virtual renunciation by Panama of intent to protect the vessel ... and hailed the dispatches as a victory of the United States in its claim of right to seize rum carriers hovering off American shores beyond the twelve-mile limit." Soon after, US Attorney George J. Hatfield advanced the opinion that the seizure was made under the provisions of a Panamanian law providing for revocation of the right of nationality of any vessel holding that country's register which was found to be habitually engaged in smuggling, piracy or illicit commerce.

With the commencement of the trial in April 1927, federal judge George M. Bourquin gave short shrift to the case, ruling that the seizure of the rum vessel *Federalship* and its million-dollar cargo and the arrest of the captain and crew were illegal and constituted "sheer aggression and trespassing like that which contributed to the war of 1812." He also pointed out that the seizure was "contrary to treaties" that the US reached with Great Britain with regard to the twelve-mile limit. He continued that "a decent respect for the opinions of mankind, national honor, harmonious relations between nations and avoidance of war require that the contract and law represented by treaties shall be scrupulously observed, held inviolate and in good faith and precisely performed, that briefly, treaties shall not be reduced to mere scraps of paper." (In his interview with Imbert Orchard, Captain Hudson mused that those attempting to enforce US Prohibition laws with *Federalship*'s seizure were never all that concerned whether the case was indeed right or wrong but were merely attempting to embarrass Consolidated financially. Hopefully, with the millions of dollars spent on lawyers and court costs, Consolidated might throw up its hands and stop rum running.)

Of course, once captain and crew were free and back aboard their ship, they were very much "up, up, aloft" according to one newspaper reporter. But still, Assistant Collector of Customs Henry E. Farmer refused to surrender the ship or release its liquor cargo until he received word from the attorney general's office. Then, on April 27, the government of Panama demanded *Federalship* be immediately released—presumably because the ship was not found guilty and was not, therefore, a pirate ship. This note passed on by the Panamanian envoy to the United States, Ricardo Joaquín Alfaro Jované, also constituted a formal protest. Among its demands was included the proviso that the steamer be re-established in the condition it was in at the time of its illegal seizure. Finally, on May 2, the *San Francisco Chronicle* in its front-page news report on the

Captain Stuart S. Stone giving a cheer as he climbs aboard *Federalship*, just after he and nineteen crewmembers were released from jail in San Francisco. Federal judge George M. Bourquin denounced the seizure of the vessel 300 miles off the Golden Gate as comparable to the causes of the War of 1812. International Newsreel, San Francisco Bureau, Photo #5690. Courtesy of Photo Morgue, San Francisco History Center Collection, San Francisco Public Library.

ship's release opened with the line, "Sailing, sailing over the bounding main." "The red tape unwound swiftly following the receipt of orders from Washington by the Coast Guard to release the rum carrier ... But where is this bonny ship and its liquid cargo, drawing away from a thirsty territory, going?" The US Coast Guard was informed that it was to turn around and convoy the rum running mother ship one hour's sailing distance from San Francisco and turn her loose to steam off on her merry way. Still, there were questions of where exactly the liquor cargo was going to end up, as well as how reparations were to be settled. Defence attorneys James O'Connor and Harold Faulkner said that it was up to the owners of the ship, but most likely one representative from each government would be appointed, who would then agree on a third neutral member to form a commission to decide on the amount of damages to be paid.

After the Coast Guard made good on fixing a broken water pipe and made other repairs to restore *Federalship* to the condition she was in at the time of seizure, and the steamer provisioned with food and water for the crew and the boilers, steam was got up. The ship was to leave San Francisco Bay under its own power instead of being towed.

PANAMA TO DECLARE
WAR ON THE USA

A particularly satirical opinion piece written by P. W. Luce appeared in "The Odd Angle" column in *The Vancouver Daily Province*'s editorial page on March 6, 1927. Luce reported that:

> Captain Stone, the hard-bitten little mariner who was in command of the rum-runner *Federalship* when she was seized by US cutters 300 miles off the Coast of California, has sworn a mighty oath that he will ask the Republic of Panama to declare war against the United States unless amends are made for treating his liquor-laden ship as a pirate ... And in the circumstances, the President will be well advised to introduce a little useful propaganda in his declaration of war. Something like this:
>
> *Senor Calvinos Coolidge,*
> *Presidente, Etats Unis.*
>
> Most Mighty Suzerain,
>
> I have been requested by one, Captain Stone, to issue a declaration of war against the United States, as a personal favor to him. As this will soothe his ruffled feelings, I have very much pleasure in notifying you, formally and by these presents, that a state of war now exists between the Republic of Panama and the Republic of United States.
>
> You will understand, I trust, that there is no personal feeling against you in this matter. My regard for you, and your good lady, remains as high as ever. I am merely carrying out my duty to my most recent subject.

As you have probably forgotten, we Panamans [sic] are of mixed Spanish, negro and Indian blood. We are mighty fighters, once roused, and I assure you that the seizure of a million-dollar cargo of our whisky is somewhat annoying, to say the least. You must remember that, in comparison to the United States, we have reached a high stage of civilization: we have no vice-president.

The cold grip of the horrid hand of war will come to your people with the spectre of slow starvation. Though it will cost us $2,000,000 in trade, we will export no more bananas to the United States! You will have no bananas!

Set that to music!

We will cut off your supply of nuts. We will deprive you of rubber. We will send no more Panama hats. We will not let you have any more ipecacuanha. And—poetic justice, since you swipe our whisky—we will sell you no more sarsaparilla!

Celebrate that in your soft drinks!

Did you imagine that, because there are 125,000,000 of you and only 428,000 of us (including Captain Stone and his merry men) you could offer an affront to our flag with impunity? Did you suppose we would tolerate the filching of our Federalship on the high seas under our registry? Did you, perchance, think we would not miss her?

I give you fair warning that we propose to seize any ship flying the Stars and Stripes that attempts to sail through the Panama Canal. We own the land on each side of this waterway for 480 miles from the Pacific to the Atlantic, and we hereby abrogate any and all treaties made with your nation regarding the internationalization of this canal.

Henceforth your fleets enter at their peril!

You will be sorry to learn that Panama is now in splendid shape to prosecute an aggressive war, up to a certain point. The government revenue last year amounted to nearly $5,000,000, and our exports were higher than ever before by $321.95. Furthermore, our railway is now fifty-three miles long, and double-tracked except in the hills, swamps and forests.

Do not flatter yourself that you have a second-rate power to deal with, even though we have no national debt. We have 130,000 head of cattle in fair shape; that is more than can be found on some of your larger ranches, and, besides, our beasts are not dehorned. We also have 25,299 horses, 660 mules and 52 asses, though, unfortunately, most of these have strayed away on the pampas at the time of writing. But for this we would be in splendid shape for immediate mobilization and there would be no difficulties whatever in fitting up the commissariat.

As the offense which has led to this declaration of war occurred off the coast of California, it is our honourable intention to make our preliminary attack on this state, which is only five times larger than the whole of Panama. After we have subdued California, and freed Captain Stone and his crew from the jail in which you propose to clap him and them, we shall be prepared to deal with you as with a conquered people. We may be generous, if you are in a position to pay a large indemnity in cash, or even in whisky.

Unfortunately, I can not set a date for the commencement of hostilities. As the independence of Panama is guaranteed by the United States, there has hitherto been no occasion for us to maintain an army or navy, a circumstance which Captain Stone seems to have overlooked in his haste to have Uncle Sam taught a salutary lesson. However, apart from this slight handicap, we are in splendid shape, so bewarrre!

On May 2, a reporter with the *Chronicle* noted with some amusement that once she was escorted out, *Federalship* would be "sailing, sailing over the boundless main." And two days later, "thumbing its nose at prohibition, the released rum vessel Federalship hoisted its Panama flag and churned out of port here yesterday afternoon with 144,000 bottles of liquor still under its belt. And now that the seagoing liquor carrier has departed speculation has risen as to what Uncle Sam's first attempt at hijacking on the high seas is likely to cost him." The other San Francisco newspaper, the *Examiner*, also had good fun with the rum ship's release in a front-page story titled "Gurgles Out with Chasers." It said that Captain Stone could be seen "standing on the bridge, puffing his inevitable cigar and grinning like the Cheshire cat."

In order to give the impression that the *Federalship* was indeed involved only in legitimate trade and thus ensure that she not lose her Panama registry by being declared a pirate ship, Captain Stone pointed *Federalship*'s bow south to presumably steam off for Buenaventura, Colombia, as declared on the ship's original manifest. Also, according to the May 6 *San Francisco Chronicle*, the cutter *Shawnee* was to "convoy" the mother ship as far as the Mexican line "to make sure the *Federalship* does not loiter near the California coast." But it wasn't over yet.

When *Federalship* arrived in Vancouver less than two weeks later (it is uncertain whether she ever went to Colombia), around forty cases of liquor were found missing. The mystery of the missing cargo was grounds for an amusing and rather sardonic editorial in a May *Chronicle* article titled "How Many Social Purposes Are There in Forty Cases?" The article noted that while "cleansed of the blot of piracy, she sailed away in dignity with her million-dollar cargo of liquor, but with a hint that some of her precious freight had fallen prey to pirates while she was in duress. Now her legal counsel speaks up and says that maybe it wasn't pirates who took the liquor, that the amount missing was after all what might reasonably have been consumed for social purposes while she was detained in port here. The amount missing is estimated at forty cases. The owners wave that aside as unimportant and a magnificent gesture. Really now, isn't this rubbing it in?"

Following up on their success with securing the release of the rum boat *Federalship*, Panama demanded the United States also release another Panamanian-registered vessel and alleged rum runner *Hakadata*, owned by Archie MacGillis's Arctic Fur Traders Exchange. The *Chronicle* reported that the US Coast Guard captured the vessel around seven miles off the Mexican coast and sixty miles south of the US border. On April 7, the 110-foot cutter *Vaughan* (originally launched as *SC-152*, a World War I subchaser) sighted a two-masted vessel lying hove to off Santo Tomas Point. Lieutenant F. L. Austin, division commander of the Coast Guard, said that when the cutter approached, the officers and crew of *Hakadata* quickly set fire to the schooner and took to the lifeboats. *Vaughan* consequently picked up the lifeboats, arrested the crew, extinguished the fire and then towed the vessel into San Pedro, California. *Hakadata* was a 57-foot, 5-inch auxiliary schooner originally launched from a Victoria shipyard in 1892 as the sealer *Saucy Lass*. Once Panama cancelled her registration—probably on the basis that her capture was legal and she was for all intents and purposes a pirate vessel— she was sold at auction that same month for $1,490.

Meanwhile, the *Federalship*, with Captain Stuart Stone in command, continued to be quite the celebrity in newspapers on both sides of the border. On June 27 she cleared Vancouver for Puerto San Jose, Guatemala, under Guatemalan registry (but still owned by Consolidated Exporters' interests). But she was never to arrive in the Central American port. Instead, she sailed in ballast to Papeete, Tahiti, where her holds were filled with liquor. The cargo had been re-routed by Consolidated Exporters in order to avoid a recently imposed Canadian transit tax on liquor transhipped from the British Isles and Europe. Captain Stone, his son claimed, was the one who actually pioneered this new means of getting liquor into the United States without having to pay a transit tax. His answer was to have the liquor-exporting country deliver up the cargo to the destination stated on the ship's clearance papers. As a result, Vancouver was bypassed and the liquor came directly from European suppliers to Tahiti. On hearing this latest development, Guatemala cancelled her registry, perhaps because the rum runners no longer had any need to bribe Guatemalan officials to create false reports about receiving the liquor, or have the vessel registered to that nation.

Regardless, after all her cargo was delivered up off the California coast in the later part of 1927, *Federalship* returned to Vancouver. After three months laid up in Burrard Inlet, with Captain Stuart Stone giving the orders, Jim Donohue back as first officer and an F. N. Eddy as second officer, *Federalship* passed through First Narrows and set out for the Netherlands. Recently rechristened *L'Aquila* (the Eagle), by Consolidated, and now under the British flag, she arrived in Amsterdam to fill her holds with liquor that, according to her paperwork, was destined for Shanghai. As was common throughout the rum running fleet, the record is sketchy as to what became of the steamer once she was out on the open ocean, and her whereabouts from May 1928 through September 1928 remained unknown to all except owners and crew. It's highly doubtful that she ever did arrive into China's port city.

Instead, in the fall of 1928, *L'Aquila* ended up at Rum Row, now situated off the Mexican coast in order to escape the ever-vigilant and bothersome US Coast Guard. Her arrival was timely, since she was on hand to rescue all hands from two other vessels involved in the rum trade: those aboard the two-masted auxiliary schooner *Jessie* and the American fireboat *Guiaicum*.

Jessie, another old sealing schooner like *Pescawha* and *Hakadata*, was owned by Arctic Fur Traders Exchange, which was incorporated back in 1922 by Archie MacGillis and Frederick Rae Anderson, a barrister. Its two shareholders and directors were none other than Archie MacGillis and William L. "Whiskers" Thompson, who had been working out of Coal Harbour soon after the Volstead Act came down. Launched as the two-masted auxiliary schooner *Ojibway* from a shipyard in Benicia, California, in 1890, the sealing vessel was 57 feet, 5 inches in length.

When researching his autobiography, Fraser Miles came across a note written by Harry Slattery, mate on the *Jessie*, dated January 8, 1929, stating that "the motor schooner *Jessie* of Vancouver, BC, was lost with the loss of ship's papers about 100 miles west of San Diego, Cal on September 23, 1928, in collision with an unknown vessel. This is made by me for Captain J. Keegan, Master of the *Jessie* who hasn't returned to Vancouver." Jim Stone said that what caused these two vessels to sink was a mystery, but most likely they either ran into each other or a US Navy warship ran them down, since the largest naval base along the west coast was right there in San Diego. Reports had it that the two boats collided after they'd taken on a load off the ss *Prezemysl,* which departed Hamburg a year earlier with 250,000 gallons of alcohol aboard. A total of 10,500 cases of liquor were said to have gone down with the two boats. The *Chronicle* explained in some detail why these boats were situated near the Mexican coast at the time of the collision: "Truckloads of Canadian liquor are being smuggled into San Francisco from the holds of rum ships in Mexican waters. The powerful Consolidated Exporters of Canada, long thought to have been beaten by the Coast Guard, has merely shifted its base of action to southern shores. Fast motor trucks and touring cars are transferring the liquor from San Pedro and Mexican ports to this city."

Fraser Miles wrote that *Jessie*'s captain, Joe Keegan, and first officer, Harry Slattery, were both good friends with Captain Stone. Miles recalled the parties hosted by his parents in their York Avenue home just above Kitsilano Beach in Vancouver, where his father's fellow mariners in the rum trade all got together in a jovial and friendly atmosphere. "These men were brothers-in-arms, knights of their particular Round Table; in the midst of their perilous venture, they were celebrating their comradeship." He remembers they were captains or, at least, mates who went on to eventually achieve captaincy: Joe Keegan, Harry Slattery, George Ford, Charlie Hudson, John Vosper, Arthur Lilly, "Whiskers" Thompson, Jim McCullough, Alec Kennedy and so on.

Following the seizure of *Federalship* and the subsequent dropping of all charges against captain and crew back in May 1927, US District Attorney George Hatfield conferred with the attorney general, whereby it was agreed that all federal departments in San Francisco were to be instructed that no more seizures of liquor-laden vessels were to be carried out more than an hour's run from American shores. But only a year and half later, this was proven not to be the case, as it appeared that the enforcers of US Prohibition laws, along with the US Coast Guard, were still dead set on getting even with their old nemesis, Captain Stuart Stone, and his command, *L'Aquila* (formerly the *Federalship*). A month and a half after their rescue of the crews of *Jessie* and *Guiaicum*, *L'Aquila* was once more making headline news following another nasty encounter with the Coast Guard.

According to a story in the November 6, 1928 *Vancouver Daily Province*, the "notorious craft" was seized after a gun battle off the Southern California coast with the US Coast Guard cutter *Tamaroa*. According to US Attorney Hatfield, *L'Aquila* had been lying off the coast for nearly three months when her provisional British registry expired on September 1. He argued that "the old offender is now a ship without a country." Both ship and captain had earned a bad reputation according to a *San Francisco Examiner* reporter: "Arriving on the west coast five years ago with a cargo of wine, dope, and women, this notorious tramp freighter has bobbed up under various names since."

According to the *Examiner*, high seas were running when the cutter *Tamaroa* approached *L'Aquila* at seven thirty a.m. on

November 5 and ran up the ensign flags recognized internationally as the mandatory injunction to stop. Of course, the ever-defiant Captain Stone, sailing under British registry and the red ensign, ignored the signal. (What Captain Stone may or may not have been aware of at the time was that a recent ruling was made by Captain D. P. A. Deotic, division commander of US Coast Guard, that gave permission to patrol vessels to fire upon a suspect vessel.) At noon, the cutter's launch approached *L'Aquila* and *Tamaroa*'s commander, Bosun William Gill, and informed Stone he had orders to board his vessel and examine its papers. When Captain Stone refused to allow the craft to come alongside, Gill replied that he would have to resort to force. Upon return to *Tamaroa*, he gave the order to fire two warning shots across the bow of the rum runner. But when she continued steaming on, a shot was laid into her broadside which stove in her "rusty black iron sheathing."

The *Vancouver Daily Province* provided a slightly different but more detailed account of the incident in its November 30, 1928, issue, after having gained access to *L'Aquila*'s "scrap" log entries (probably the logbooks thrown overboard if about to be boarded by the Coast Guard), penned by Mates Donohue and Eddy. They reported that on November 5, the old steamer was 121 miles from US waters when the 151-foot-long *Tamaroa* (a tugboat so slow it was called the "sea cow") ordered her to "heave-to" and prepare for a boarding party to examine *L'Aquila*'s papers. Instead, in a repeat performance of his response when *Federalship* was seized off San Francisco back in March 1927, Captain Stone grabbed a megaphone to shout across the water "Stop, what for? I'm on the high seas!" Once again, a US Coast Guard cutter opened up with her gunnery. As Hugh Garling, who was later to sail under Stone aboard *Malahat*, pointed out, "Captain Stuart Stone was long on nerve, and this served him well during his service in the rum running fleet."

The *Daily Province* continued, "The cutter opened fire with solid shot, sending six shells, four of which took effect in the hull just forward of the rudder, and above the waterline ... Captain Stone then hove-to and was boarded by the commander of the *Tamaroa*, who examined his papers and then informed Stone he would have to wait orders from his headquarters." Once a prize party took control of *L'Aquila* (her red ensign was hauled down amid strong protests from her crew) and Captain Stone was removed to *Tamaroa*, his ship was put on the towline for San Francisco with the cutter *Cahokia* standing by. But they never arrived, much to the distress of Coast Guard officials, since an order was received by radio to release the ship. As it happened, government attorneys discovered that they'd made a serious blunder: the ship's provisional registry hadn't expired in September after all, since a clause in the ship's papers declared her British registry was for a six-months period or "until reaching a British port." As a consequence, Captain Stone was allowed to return to his ship, where he promptly got up steam and headed for Mexican waters. The *San Francisco Examiner*, reporting on the ship's release on November 8, pointed out that "since leaving Amsterdam for Shanghai, with a $1,000,000 liquor cargo, the Federalship [*L'Aquila*] has not touched at a British port. About $100,000 worth of liquor is left on board it is believed."

While the crisis seemed resolved and the ship's papers returned at two p.m. on November 7, Hugh Garling said that *Cahokia* persisted in harassing *L'Aquila*. At four p.m. her commander informed Captain Stone he now had orders to take her in tow. An hour later, as she approached the Mexican three-mile limit, *Cahokia* overhauled *L'Aquila* and gave four blasts of her whistle, the order to stop, two times. Running close aboard, the master of *Cahokia* hailed Stone: "If you do not stop, I will open fire upon your ship." Stone's reply? "I hereby warn you that you would be firing upon the British flag on the high seas and if you kill any of my men, I will hold the United States responsible!" Even though eight shots were fired across his bow, Stone continued to steam on as *Cahokia* ceased firing and dropped astern. But then, early in the morning on November 10,

Tamaroa returned and approached, and her commander demanded *L'Aquila* stop. But with the Mexican shoreline looming ahead in the dark, Captain Stone semaphored the master of *Tamaroa* that he now considered his ship safely within the Mexican three-mile limit and was soon steaming into Ensenada harbour.

After he got back to BC waters, Captain Stone filled a *Vancouver Daily Province* reporter in on the behind-the-scenes efforts that were made to hand them over to the Americans once *L'Aquila* was set free. "The US Consul at Ensenada ... endeavoured to prevent the entry of the ship into the harbour, but the Mexicans, in view of the fact that shells from the Tamaroa had pierced the fresh water tanks of L'Aquila, permitted the ship's right to sanctuary." The ss *Kuyakuzmt* (previously *Stadacona*) arrived in Ensenada and removed the ship's cargo before taking her crew back to Vancouver later that November. It is interesting to note that Joseph Hobbs, the steamer's owner, christened her with the unpronounceable name of a BC mountain, Kuyakuz, and the ship's painter probably decided the name was supposed to be "Kuyakuz Mt." and inscribed that on her bow. As Johnny Schnarr said, rum runners often chose complex names for their vessels in order to make it harder for the Coast Guard to identify and put a name to them. Sometime shortly thereafter, the rusty remains of *L'Aquila* were found sitting in Ensenada harbour, where she was slated to be converted into a fish cannery.

Her entries in Lloyd's over the next two years provide a good log of what became of the old steamer. Once she was registered in Ensenada, Mexico, she underwent two more name changes under new owners. In 1934, she was listed as *Calmex*, and then in 1935, while owned by a Carlos E. Bernstein, was renamed *Marmex* and was apparently still in existence as such up until 1945. In his book, *Rum War at Sea*, Commander Malcolm F. Willoughby wrote that the retired mother ship eventually ended up back in British Columbia coastal waters by the early 1940s, working under the name of *La Golondrina*. Here she was put to work as a tug towing booms of logs between the logging camps around Haro Strait (probably Hecate Strait) and Prince Rupert. He also noted that in October 1940, she responded to the distress signals of the steamship *Alaska*, which was stranded in BC waters between Ketchikan, Alaska and Prince Rupert. The other vessel on hand to help refloat *Alaska* was the US Coast Guard cutter *Cyane*, whose commander, Lieutenant Frank W. Johnson, back when he was gunnery officer on *Cahokia* fired the shot that hit *La Golondrina* when she was still operating as *L'Aquila* back in November 1928 (this Frank Johnson was most likely the same man described as "Frank K. Johnson" who was ordered to fire on the *Federalship* in March 1927).

Even though it was sensational at the time and made good press, Fraser Miles pointed out that the seizure of *Federalship* in 1927 and then again as *L'Aquila* in 1928 was not the primary cause of the rum runners' poor showing in what he termed the early Coast Guard years. He implied that the real cause was poor management and a lack of good organization. Miles states that what the captures did succeed in accomplishing was to spur "the rum ships' owners to either replace their masterminds, or to draw up new plans to get move the whole operation out of the doldrums." As it happened, a well-bred British gentleman with all the right qualifications just happened to become available to get the whole complex operation back on track.

Chapter Ten

GETTING ORGANIZED BIG TIME

CAPPY HUDSON TAKES OVER (1928)

While Archibald MacGillis was considered one of the founding fathers of West Coast rum running after he jumped into the game early managing a couple of liquor export enterprises out of Coal Harbour, by the late 1920s it was Captain Charles H. Hudson, or "Cappy" Hudson, as he became known, who was recognized as its mastermind. Until Hudson arrived on the scene, the liquor export business was all rather loosely run, and it wasn't unusual for someone like Archie MacGillis, who established a very profitable arrangement with Consolidated Exporters early on, to show up at their head office and warehouse on Hamilton Street in Vancouver to ask for fifteen thousand dollars to buy stores and supplies for his ships. With Hudson at the wheel, his wealth of knowledge combined with his finely honed organizational skills transformed the company—with its access to a fleet of some sixty vessels and an overly complicated system of legal manoeuvring, transport and communication—into a sophisticated marine transport operation. Under Hudson's command the enterprise proved very profitable and especially rewarding for its directors and shareholders.

Rum runner Hugh Garling recalled meeting Hudson in the early 1930s when he applied for a job as a seaman. "He was the antithesis of the commonly held concept of a rum runner. He was a

slim distinguished-looking gentleman and seemed to have all the charm of a well-bred gentleman. I was seeing the velvet glove covering the mailed fist, for he could be tough if need be." Jurgen Hesse, who interviewed Captain Hudson for a *Western Living* magazine story in 1976, described him as having "patrician looks, gentle humour and a courtly bearing." Hudson was of English birth and descended from five generations of mariners. He ran away to sea when he was fourteen years old and came of age to serve with the Royal Navy during World War I. He distinguished himself by being twice awarded by the king. He received the Distinguished Service Cross (DSC) in September 1915 for his services in the submarine commanded by Lieutenant-Commander Cochrane in the Sea of Marmara and then a bar for the DSC for service in action with enemy submarines. By this time, he was in command of HMS *Gunner*, a Q-ship or mystery ship, a new trawler that was one of the thinly disguised armed merchant vessels used for hunting submarines.

The velvet glove covering the mailed fist. The distinguished-looking and ever-charming Captain Charles H. Hudson. Photo published in Hugh Garling, "Rum Running on the West Coast: A Look at the Vessels and People," *Harbour & Shipping* magazine, July 1989.

Following the end of the Great War, Hudson immigrated to Canada, where he and his brother made a futile attempt at farming in Manitoba. After four years of struggling and faced with an accumulated debt of some ten thousand dollars, Hudson decided to head for Vancouver to take up the seafaring life once again. In 1923, Winnipeg's chief of police, an acquaintance of Hudson's, gave him three letters of reference to help him find a new career. One was addressed to a "big politician"; the second to another demobilized Royal Navy officer, Barney Johnson, owner of Hecate Straits Towing and the shipping and insurance agency B. L. Johnson and Company; and the third was addressed to a Captain William L. Thompson. Hudson landed his first job when he walked into the Canadian Pacific Railway coastal service office and got on with the steam tug *Nanoose*, where he was paid $138 a month as a mate, but once he realized he was never going to pay off his debt with this paltry wage, he was eager to jump at any opportunity to earn a better dollar. So he walked down to Coal Harbour and, with reference letter in hand, introduced himself to Captain Thompson— who Hudson's wife nicknamed "Whiskers," and who Hudson recalled was "awfully English"—down at his 38-foot fishboat. When Hudson asked what he was using it for, he replied, "rum running." And when he then asked what that might that involve, Thompson replied, "We pick up booze and take it down to the States and sell it and make money."

With Hudson showing some interest, Whiskers suggested that maybe he'd like to take charge of the old two-masted sealing schooner *Borealis*, 71 feet, 6 inches in length, that very night. In an article that appeared in *Harbour & Shipping* in May 1964, Ruth Greene wrote that Hudson told her that after leaving Vancouver in ballast, they sat outside Cape Flattery where they met a "parent ship" which was loaded with choice booze. Then after loading some thousand cases aboard, they proceeded to an arranged rendezvous with an American boat to which they transferred their cargo. Hudson had successfully pulled off his very first run in the liquor export trade. He was delighted to tell Greene that he "made $400 for a day-and-a-half's work. This was after losing money for years on the farm and working like a slave! What a change to receive $400 for six days' pleasure!"

Not only that, once he got back to Coal Harbour after a successful six-day voyage running down to the California coast to deliver up liquor, he received a five-hundred-dollar bonus.

After a couple more runs with *Borealis*, Hudson took charge of Archie MacGillis's three-masted auxiliary schooner *Coal Harbour*. Now he was involved in the big time, making runs transporting some ten thousand cases of liquor aboard *Coal Harbour* to sit outside US territorial waters and wait patiently for American buyers' boats to come alongside. At this time early on in the game, Rum Row was still situated off the Farallon Islands outside San Francisco Bay.

Once the court case following the capture of the *Coal Harbour*, and the subsequent trial of Hudson, his officers and crew was settled in June 1928, Hudson was back in Vancouver, where he was taken on as marine superintendent manager, or "shore captain," of Consolidated Exporters. He stayed with the company as its head for the next twelve years and right from the start was "very, very keen because they didn't know their foot from their elbow ... None of them understood a damn thing about boats, so when I got in it was real terrific to handle things and get in front of them and tell 'em what I had to do or didn't have to do and explain it to them ... and [they] eventually let me have my own way. I had wonderful time ... a Capt. McLellan came along, supposed to be my boss, but didn't know foot from elbow. I had to run the whole ruddy works. I'd come back with, like I say, $60,000 and throw it on the desk ... They'd ring me up at 2:00 in the morning and give me an order a mile long which I put down to the boats."

Captain Hudson remained quite proud of his role in rum running during the Prohibition era and even argued that during its heyday the trade was the economic saviour of the city of Vancouver. As he told oral historian Imbert Orchard in a taped 1960s interview for CBC Radio, "Vancouver was in the midst of a real depression, with logging, fishing, mining, etc. in the doldrums. It took rum running to keep industry going, especially on the waterfront. The tremendous moneys paid out to industry in Vancouver were never known to the average citizen. We spent a fabulous amount of money building boats, purchasing and overhauling engines, buying food

592M

With her long, narrow hull, *592M* was built to negotiate rough seas at high speed. Although hers wasn't a planing hull, it was claimed she was still capable of thirty knots in the roughest of conditions out in the Strait of Juan de Fuca while carrying a 3,000-pound payload. In order to make her almost impossible to spot out on the water, the boat was painted in a grey-green flat finish that designer Leigh H. Coolidge called "no-see-um." *592M* had an armoured superstructure and watertight bulkheads to protect her crew and cargo. Museum of History & Industry, Webster & Stevens collection, #1983.10.11952.

and supplies for our ships, using the shipyards for overhaul and in wages for the crew and fuel. We had a Japanese yard down by the [Rogers] sugar refinery and then we had them over in Vancouver ... All the big shots with money didn't know a darn thing how to handle stuff, so I was purchasing agent. I bought all their fuel, all their food, doctor's supplies; I hired every man that went on the ruddy boats and superintended all repairs ... and I did that for $450 a month!"

Since the whole operation was to all accounts a bona fide international shipping enterprise, it required skilled and sophisticated supervision and management and Captain Hudson was the ideal person to be in charge when he took it over in 1928. Captain Hudson not only had the knowledge acquired from his years commanding rum ships, but soon showed a natural talent at managing a large and complex business enterprise.

Cappy Hudson proved to be Vancouver's equivalent of Seattle's Roy Olmstead, who was a sophisticated, charming and astute businessman who abhorred violence. As well as proving himself to be an astute logician, Hudson excelled at customer relations since, according to Hugh Garling, he was able to engender a strong feeling of trust and confidence in his buyers. "He spoke of them as charming people, none of them tough thugs or racketeers. When they came up, he'd meet them and take them home, perhaps for supper. Often, they'd give him, for instance, $60,000 for their order and he'd put it in a box on the mantelpiece, or sometimes, he'd wrap it up in a newspaper and throw it on the floor of the car ... The buyers trusted him implicitly and his word was his bond. Likewise, he worked with them as gentleman to gentleman." Throughout Imbert Orchard's interview, Hudson confirmed Garling's assessment, and he comes across as very professional as well as one very self-assured individual. As writer Ruth Greene described him, he was, like Roy Olmstead, adventure-loving, gentlemanly and a real man of action. In one of his oral tapes, held in the BC Archives, Hudson was quite open about what it was like in his new role working with Consolidated. "We had a wonderful, wonderful time! ... The people down below, our [American] customers, trusted me implicitly; they would ring me up, 'Look, Cappy, will you release so and so?' ... 'Well, shoot me up $30,000' ... 'It will be there in the morning.' And they never failed ... Got there in the office in the morning and I threw $62,000 under the clerk's desk, 'That's so and so's money' ... go into the bank anytime and cash their cheques since their cheques were always good."

Even later in life, Hudson remained steadfastly discreet. For instance, when he was interviewed by Imbert Orchard, he seldom, if ever, mentioned any of those he had dealings with throughout the rum running years by name. When he did speak of a particular individual involved in the trade, or who had business with Consolidated Exporters, he only referred to them as "so and so," which it made particularly difficult for researchers and historians to sort out the business relationships of those involved in the trade long after the fact. Also, as Max Henderson, who worked for Sam Bronfman before becoming Canada's auditor general, recalled in his memoirs, "Extreme care was exercised; not to commit anything but the absolute minimum to paper while ensuring that the transactions were strictly legal."

Captain Hudson also emphasized during his interviews that "there was two Vancouver distilleries going full blast, business was flourishing ... the mother ships were back and forth ... There was the *Malahat, Coal Harbour, Stadacona, Mogul, Hurry On, Quadra* ... and all the smaller boats, I could give you another 20, 30, 40, 50, and all these boats they kept Vancouver going! People might criticize it, but it was God's gift to Vancouver at the time ... which in 1922 was in the midst of a real depression, with logging, fishing, mining in the doldrums. It took rumrunning to keep industry going, especially on the waterfront ... We were the only ones building boats ... and Clarey Wallace, boss of the big shipyard in Vancouver, came to me at my office. Why didn't he get such and such a boat to build? That's how tough it was for them. We brought prosperity back to the harbour of Vancouver."

Under Hudson's capable direction the whole operation became a smooth-running machine but, like Roy Olmstead's Seattle-based operation, probably the biggest contributor to its success was the establishment of a reliable ship-to-shore communication system early on. Wireless radios played the key role in keeping track of the locations of the various vessels in their fleet, since it was essential

The schooner *Malahat* lying alongside the Balfour-Guthrie Warehouse, just east of Ballantyne Pier where a steady stream of liquor consignments arrived in Vancouver harbour. Cargo aboard these various freighters, liners and steamships from Glasgow, London and Antwerp was unloaded and put into Federal Department of Customs and Excise security. City of Vancouver Archives, CVA 447-2426, photo by Walter E. Frost.

they maintain contact with the home base. Often positions had to be changed in a hurry and Vancouver alerted in the nick of time.

In one of his oral interviews with Ron Burton in the late 1960s, Hudson continued to explain: "The biggest boosters wireless ever had was the rum runners." Hudson knew of a ham radio wizard, Bruce Chisholm, with Sprott Shaw in Vancouver. After he approached Chisholm and inquired if he was up on all this "short wave radio stuff," Chisholm replied he was, and even had two or three up and running already. "We could reach Tahiti from Vancouver, so we put all our boats on shortwave, and soon all the boats on the coast had them. Ship to shore phones, we had the first, I started the whole thing. We had our own set up in the East End of Vancouver." Captain Hudson also installed radio equipment in his Vancouver home in Point Grey in order to keep in touch with the ships at sea belonging to or chartered by Consolidated Exporters

2209 YORK STREET, VANCOUVER, Canada

193........

Radio.................. Ur Sigs Wkd hr at..................Gmt

Note..........Qsb.........Qsa........Qrm........Qrn..........

CAPT.
STUART S.
STONE

VE5AF

MISS
H. M.
STONE
(HAZEL)

XMTR..................................Recvr..........................

Remarks..

Tri Agn O.M. ....OPR

Captain Stuart S. Stone and Hazel Stone's shortwave, ham radio licence. Fraser Miles collection.

while providing them with nightly communiqués with the latest news regarding the positions of US Coast Guard cutters.

Much like the internet and smartphones of today, wireless telegraphy—or radio as it came to be called—was leading-edge communication technology in the 1920s. Up until that time, the telegraph and then the telephone had been the most effective means of communicating with distant locations. Once the transmission of signals through free space by electromagnetic radiation or radio waves was discovered, a physical wire connection was no longer required. As Jim Stone said, it was indeed the rum runners who introduced state-of-the-art high frequency radio equipment—still dot and dash Morse code communication for the most part, but also radio telephone.

As it happened, Captain Stuart Stone, the infamous captain of mother ships, had a sister, Hazel, who lived with Stone, his wife and their children. At the time, Hazel was working as an operator at the BC Telephone exchange downtown on Seymour Street, and was taking wireless lessons at Sprott Shaw so she could learn to operate the radio equipment in Captain Hudson's home. Stuart's son, Jim, said in the biography of his father that "every evening she took the Fourth Avenue streetcar to the western end of the line and made her way to Hudson's Point Grey residence where she spent several hours (usually between 9 and 11 p.m.) transmitting instructions to the rumrunning fleet, designating a rendezvous point, the identity of the contact boat and the cargo to be passed over, as well as other pertinent details." As Hudson continued to explain it, "We used a slaters' code ... that anybody could decode. I was the Marine Superintendent Manager, so I made up a brief code, possibly one page of 'I am on position/not' 'Weather is good/bad' etc. It all began with 'B.' We had a Chinaman in the East End of Vancouver ... said he had a message for us, the guy running a small boat straight into shore so I wired the buyer, "mother in hospital," he picked up the booze, this was the first shortwave message received in Vancouver, a thousand cases of champagne and nobody knew a thing about it ... It all worked like a clock."

The new radios, courtesy of Sprott Shaw, were far more effective and reliable than the original ones—but they were also so compact that it was easy to take one down and hide if an inspector happened to step aboard. Once they had the radios, Captain Hudson also looked to Sprott Shaw for operators of their shortwave receivers and transmitters. Many of the boys were taken on by the rum runners, once they had learned to operate a set. These were students who agreed to sail aboard ships loaded with unknown cargos to unknown destinations for two hundred and fifty to three hundred dollars a month.

Young Fraser Miles, who grew up in Mission in the Fraser Valley and was nineteen years of age, just out of high school, happened to come across an ad in the help section of a Vancouver paper in September 1930: "Youth wanted for half day work, live in for free board, room and wireless telegraphy course." After he was interviewed by Mr. Sprott himself, who he described as a fat and very deaf guy, he accepted the job and moved into Sprott's home on

Marine Drive and was soon on his way to becoming a wireless operator. He lived in the basement of their home and in exchange for his wireless education, took care of their large garden, tended their sawdust-burning furnace and washed and polished their cars.

In his autobiography, Miles recounted how he was still in Vancouver over a year later, and "hadn't earned a nickel in all that time," when he was asked by a buddy to help load stores on a boat down at the Evans, Coleman and Evans dock in Burrard Inlet in December 1931. It was the packer *Ruth B*, which appeared somewhat odd to him since its decks and paintwork were so clean, with not a spot of rust on her ironwork. When the wireless operator didn't show up that evening, and Miles let the crew know he'd just spent a year at Sprott Shaw, where students practised day after day on a wireless, he was hired as a replacement. All he knew was, "I had a job on a boat, a boat called the *Ruth B*, a fish packer that hadn't carried fish in years. That was all I knew or could guess. What I didn't know didn't seem important, like what the boat carried, where it was going, when it would be back, the name of the company, and not least, what I was being paid for my as yet unknown duties ... Rum runners, as a group, made the Sphinx sound like a chatterbox ..." since it didn't take long to catch on from tight-lipped fellow crew members that "the less anyone knew about our other activities the better, and I followed instinctively the rum runner's motto: 'Don't never tell nobody nothing, nohow.'"

ACROSS THE LINE, FAST AND DIRTY

THE PURPOSE-BUILT RUNNERS (1928-33)

"Hoozit Chained Up by Customs" a bold headline declared on *The Daily Colonist*'s Marine and Transportation page on October 14, 1928, "Her owners failing to produce a fine of $300 inflicted by the Department of National Revenue, the Tacoma speedboat *Hoozit* is now padlocked to the *Miss Victoria II* at the Causeway Boathouse. The two men found aboard the vessel when she was overhauled by the *Despatcher* on Friday morning, J. Rice and Louis Bussanich, were fined $20 each by immigration authorities for being in Canadian waters without reporting themselves. One of the neatest speedboats brought to Victoria by Customs officers, the *Hoozit* is double planked with a small but heavily built pilot house, with extra-thick plate glass windows. The symbol Net 5, No. 226969, is carved in large letters on the starboard side of the cabin wall. The *Miss Victoria II*, to which the *Hoozit* is chained, is tied with fifteen feet of rusty chain and two padlocks, has been held for investigation here since March 9. She was seized following a visit to this port of one of the Port Townsend Coast Guard cutters, whose crew claimed to have fired on her in a sensational chase." Louis Bussanich, the master of *Hoozit*, was a blacksmith from Dockton, Washington, while his partner, John Rice, who Hugh Garling described as a tough customer of Scottish descent and also a Dockton resident, was an automobile mechanic.

The two captured boats tied side by side just below the Empress Hotel and Parliament Buildings soon proved quite the attraction. The purpose-built rum runner *Miss Victoria II*, owned by local resident Johnny Schnarr, was a regular visitor to the Inner Harbour and could often be seen loading liquor down at Wharf Street below the Consolidated Exporters' warehouse at the bottom of Fort Street. *Hoozit*, on the other hand, was an American-owned rum runner and would have been a rare sight indeed. Like her Canadian counterpart, she was built to a very fast design and, according to Hugh Garling, was 40 feet in length and powered by a four-hundred-horsepower engine capable of driving her at around thirty-two knots. Two days earlier, on October 12, the Canadian customs vessel *Despatcher* had pursued and captured *Hoozit* off Beaver Point, Salt Spring Island.

In his report to J. C. Barton, divisional chief of the Customs-Excise Preventive Service in Vancouver, Officer G. E. Norris, assistant inspector in Victoria, related the circumstances connected with the successful capture of the American runner. "By steering a course among the rocks south of D'Arcy Island, the *Despatcher* headed the *Hoozit* off from reaching American waters ... When we were off Salt Spring Island, the *Despatcher* was to the eastward of the *Hoozit*, darkness was upon us; the sun had set; but the light remaining in the western sky had placed the *Hoozit* in comparative light and the *Despatcher* in the dark. The *Hoozit* evidently having lost sight of us came up to anchor and lay sheered against the steep shore ... In this position the *Hoozit*, which had no lights burning could not have ordinarily have been observed by passing vessels but would have merged into the shoreline."

This particular capture was especially rewarding for Canadian customs working on the Pacific coast at the time, since *Despatcher* had only recently slid down the ways of a local shipyard. By the mid-1920s, Canada's Customs and Excise Department was frustrated with being continually hampered by inadequate resources in its attempt to enforce customs regulations and, in particular, combat the traffic in liquor contraband by sea. In order to deal with the proliferation of fast purpose-built rum runners throughout local waters, in November 1927 the department awarded a contract to S. R. Wallace of North Vancouver for the construction of two express cruiser patrol boats, *Imperator* and *Despatcher*. In the September 1991 issue of *Harbour & Shipping*, Hugh Garling went into some detail describing how the *Despatcher* was designed to be one of two of the fastest boats ever constructed for the Pacific Command of Canadian Customs and Excise. Built to the designs of H. S. Hoffar, the larger of the two, *Despatcher* was 49 feet overall with a 10-foot, 4-inch beam and moulded depth of 4 feet, 10 inches. According to Hugh Garling, Hoffar guaranteed a speed of thirty knots or more from twin "Dolphin Special" Sterling gas engines of 290 horsepower, which turned her twin twenty-two-inch by twenty-two-inch, three-bladed propellers at 1,950 rpm. A tripod mounting for a machine gun served as a mark of her authority. Garling didn't provide any details about the smaller *Imperator*.

There were still countless men and boats immersed in the coastal rum running trade who continued to earn a decent, and often very lucrative, living. Perhaps their biggest key to success was having the skill and ability to elude enforcement forces and hijackers on both sides of the line. As Hugh Garling noted, "For every boat seized, a hundred landed in safety. For every dollar's worth of liquor seized, the smuggled liquor reaped $1,000 from successful sales, and for every man arrested hundreds more ignored the law."

Archie MacGillis caught on early that to succeed in the trade, fast boats were essential. In 1922, MacGillis picked up two retired US Navy subchasers in Puget Sound, the *SC 293* and *SC 310*, which had both seen service in World War I, and after renaming them *Etta Mac* and *Trucilla* respectively, he initially put them to work with his towing company. But since the subchasers were both fast and hard to spot out on the water, he soon had them delivering liquor directly into US waters. For some reason, however, he returned *Trucilla* to

Built as US Navy subchasers, *sc 293*, Archie MacGillis's *Etta Mac*, and *sc 310*, *Trucilla*, had long, narrow hulls that made them ideally suited for their new role working as rum runners. Sacks of liquor are being loaded aboard *Trucilla* via a canvas chute from what is most likely the Ballantyne Pier, Burrard Inlet, Vancouver. City of Vancouver Archives, CVA 1139, photo by Stuart Thomson.

the company's more mundane barge-towing service around 1924. (There's no record of when *Etta Mac* returned to towing barges.)

Trucilla was built in 1918 as *sc 310* at the Puget Sound Naval Shipyard in Bremerton, Washington. At 105 feet, 4 inches in length, she was one of two subchasers sent to Alaska during World War I. The wood-hulled subchasers were launched with three six-cylinder 220-horsepower Standard Motor engines that were both enormous and hard to maintain, along with one Standard Motor two-cylinder auxiliary engine. Once in MacGillis's hands though, *Trucilla* was repowered with a single sixteen-cylinder Standard Motor diesel.

A Liberty v-12 engine. US Air Force photo.

Etta Mac was launched from the same Navy yard in 1918 and was 94 feet, 6 inches in length. MacGillis had her three Superior engines replaced with three 450-horsepower Liberty L-12 gas engines. These surplus World War I fighter plane engines were the rum runners' engine of choice. By taking advantage of the subchasers' fast hulls and increasing their power output, MacGillis most likely stimulated others, eager to get going in the trade, to either get their hands on a chaser or to at least design and build a fast hull of their own.

After the United States declared war on Germany in April 1917, the Liberty engines were being built to power an aerial armada of fighters, observation planes and bombers being constructed for a planned 1918 spring offensive. The first run of Liberty L-12s were four-hundred-horsepower, water-cooled v-12s designed for a high power-to-weight ratio and ease of mass production. Victoria rum runner Johnny Schnarr, who had five fast launches built during the Prohibition years, noted "the Liberty engines were selling through [US] boating magazines [and] were not really being used in airplanes anymore. Somebody was buying them up cheap, making the necessary adaptations for boats, and then selling them for a hefty price to rumrunners."

The Reifel family of brewers and distillers, through one of their liquor shipping arms, the Atlantic & Pacific Navigation Company, also happened to have a former US Navy subchaser among their rum running fleet: *sc 308*, which was built in the US Naval Shipyard in Bremerton, Washington, probably in 1917. She was 94 feet, 10 inches in length and, according to a US Coast Guard report, her three Standard engines were also fitted out with smoke-making apparatus for protection if a Coast Guard cutter caught sight of them. Once she was in the hands of the Reifel interests and entered in the Vancouver Ship Registry in 1930 as a "cargo carrier," she was renamed *Hurry Home*. They then had Vancouver Shipyards remove her centre Liberty engine (the presence of the Liberty engine indicates that at least one of her engines had been replaced at some point and she was most likely already being used as a Yankee rum runner before being bought by the Reifels) and install a new Cummins Diesel, "to secure increased economy and greater cruising range at moderate speed," according to *Harbour & Shipping* magazine's November 1930 issue.

Two other subchasers bought by the Reifels were transformed into the "cargo carriers" *Ragna* and *Zip*. *Ragna* (previously *Ramona* of Mexican registry) was registered in Vancouver in January 1930 as owned by Pacific and Foreign Navigation Company, a Reifel shipping outfit operating out of Vancouver. *Ragna*'s entry notes that she was 94 feet, 6 inches and was launched from the Mare Island Naval Shipyard at Vallejo, California, in 1917. She was powered by a two-hundred-horsepower diesel along with a 450-horsepower Liberty gas engine. Later, in December that same year, the shore boat's registration was transferred to the Reifels' Atlantic & Pacific Navigation Company, located at the same address as the Pacific and Foreign Navigation Company. The second subchaser bought by

Zip loading in rough weather alongside the mother ship *Lillehorn*, which is lying at anchor on Rum Row, Ensenada. Fraser Miles collection.

Atlantic & Pacific Navigation was renamed *Zip*. The retired US Navy vessel was 109 feet in length and launched in 1918 from the Puget Sound Naval shipyard. Following their purchase of the vessel, the Reifels had Vancouver Shipyards recondition *Zip* and install a new 140-horsepower Atlas-Imperial diesel to turn the centre prop and two 450-horsepower Liberty gas engines to drive her wing props.

The other big-time liquor export shipping firm was the General Navigation Company of Canada, another shipping subsidiary of Consolidated Exporters, which was operated out of downtown Vancouver. Its principal shareholders upon incorporation on July 28, 1931, were "Charles Henry Hudson, Master Mariner and John James Randall," both of Vancouver. Among its fleet was the *Hickey* (previously the *Ocelot*). This particular subchaser-to–rum runner conversion was originally built for the US Navy in San Francisco in 1917. It was 110 feet long and powered by a six-cylinder Cooper Bessemer 150 horsepower diesel and, of course, two Liberty twelve-cylinder gas engines capable of delivering four hundred horsepower each.

During the twilight years of US Prohibition, Burrard Inlet shipyards were kept well engaged building fast shore boats, which

Another subchaser/rum runner conversion, *Hickey* (previously *Ocelot*) was originally built for the US Navy in San Francisco in 1917. Fraser Miles collection.

were recognized as "750-case boats" since that was their carrying capacity. In 1929, Harbour Boat Builders Limited (managed by A. Hisaoka), was located down on Powell Street near the Rogers family's BC Sugar refinery in Burrard Inlet. They built a number of "fast yachts." Mr. Hisaoka's company was one of many BC boat builders staffed by skilled Nikkei workers (people of Japanese ancestry), who were already recognized for their excellent fish-boats. By the late 1920s, they were also keeping their crews busy building fast purpose-built rum runners. These included *Pleasure*, built for Vestad Estates Limited of Vancouver (the Vancouver head-quarters of the Reifel interests) in 1928. She was 60 feet, 8 inches in length and powered by two 580-horsepower Sterling engines. The following year, *Yurinohana*, 71 feet, 11 inches in length, slid down their ways. Powered by one 200-horsepower Atlas-Imperial diesel and two 450-horsepower Liberty engines, the shore boat's owner was listed as E. G. Woodside of Vancouver. *Corozal* followed soon after. At seventy-two feet, two inches in length, the "cargo carrier" was powered by a "foreign full diesel engine; BHP 300" according to *Harbour & Shipping* in February 1930. It was owned by General Navigation Company (a subsidiary of Consolidated Exporters).

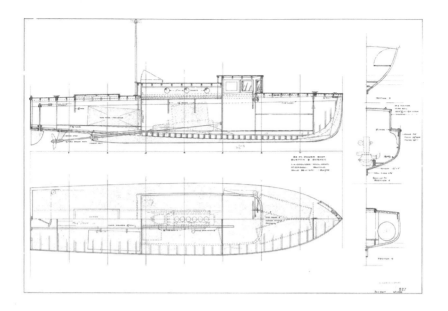

The ship's plan for the purpose-built rum runner *Pleasure*. Published in Hugh Garling, "Rum Running on the West Coast: A Look at the Vessels and People," *Harbour & Shipping* magazine, October 1991.

Meanwhile, Bidwell Boat Works in Coal Harbour, whose owners were listed as the Nakamoto Brothers, was also kept busy building high-speed freighters recognizable by their typical long, narrow hulls. *Amigo*, at 55 feet, 5 inches in length, was launched in 1931 with one six-cylinder, 150-horsepower Cummins Diesel and two 450-horsepower Liberty engines. The owner was listed as Triangle Freighters of Vancouver. Then in 1932, the 47-foot, 5-inch *Adanesne* ("Ensenada" spelled backwards) was built for Arctic Fur Traders Exchange, whose directors were listed as Archibald MacGillis and William L. Thompson. She was powered by one six-cylinder Buda 125-horsepower diesel and two 450-horsepower Liberty gas engines. That same year, the 54-foot *Colnet* was launched with an eighty-horsepower Atlas-Imperial and a single 450-horsepower Liberty engine. It was built for Consolidated Exporters' subsidiary, General Navigation. Also, in August of that same year, *Harbour & Shipping* reported that another 56-foot freighter was being laid

Yurinohana in Vancouver harbour. Fraser Miles collection.

down while both *Adanesne* and *Colnet*, which had just been completed, left port.

Finally, *Yukatrivol*, launched from the Coal Harbour yard of Union Boat Works early in 1933, had the distinction of being the last purpose-built rum runner on Canada's West Coast. She was 59 feet, 9 inches in length, and three 450-horsepower Liberty engines turned her screws. Her owner was listed as Arctic Fur Traders Exchange. As it happened, *Yukatrivol*'s first and only voyage down to Rum Row off Ensenada was from mid-April 1933 through to the end of that year, since Prohibition was all over and done with by that December. Regardless of her short career in the liquor trade, the fast cruiser is still afloat today as the elegant yacht *Sea Ox* and is often seen around British Columbia waters.

One individual who jumped in early to design and build his own purpose-built fast shore boats was Victoria-based mariner Johnny Schnarr. With his coastal logging and fishing background from growing up in remote Bute Inlet up the coast from Vancouver, Schnarr had a wealth of experience out on the water and had also developed two particular skills that had a big influence on his becoming one of the most successful—and wealthiest—rum runners along Canada's West Coast. To start with, he had a knack for things mechanical, so was quite handy with engines even before he entered the trade. And once he became immersed in rum running he revealed a natural talent for designing high-speed wood hulls. Another factor contributing to his success was his willingness to assume a high amount of risk jumpin' the line to deliver up his cargo directly onto Washington state beaches.

Schnarr said he never had any qualms about becoming involved in the dicey liquor trade, since it was believed that the law would eventually be repealed anyway. He started out by hauling liquor in November 1920 in the 28-foot *Rose Marie* owned by a fellow named Harry who offered to pay him five hundred dollars to pilot his boat down the coast to San Francisco. Unfortunately, they ended up wrecking on a sandbar at the mouth of the Columbia and had their cargo stolen by locals. After they returned to Victoria, Schnarr went to work hauling for Fred Kohse, who was living in a boathouse down in Victoria harbour. Schnarr and his partner would load liquor from a wharf conveniently located just below Consolidated Exporters' warehouse in the Pither & Leiser building on the corner of Fort and Wharf Streets in Victoria's Inner Harbour. (This building still has the iron bars on its street-level windows to protect the thousands of cases of liquor that were once stockpiled in its basement.) American importers would first visit Victoria, place their order at Consolidated and then have runners like Schnarr freight it out to their American counterparts. And it was only a matter of visiting the local customs office for clearance papers to keep it all legal and above-board. Schnarr said that the Canadian government had no problem with the export trade into the US in those early days, since they were more than happy to collect their nineteen dollars in duty per case of liquor.

The Pither & Leiser building was ideally located for loading boats. It was situated on Wharf Street, just above a loading dock on Victoria's Inner Harbour. Courtesy of City of Victoria Archives, CVA M08578.

Leaving Victoria Harbour with seventy-five cases in the small boat, Schnarr's routine was to time it so that they would be crossing Haro Strait in the dark in order to meet their contact on a beach just a little north of Anacortes, Washington, that same night. After pulling off five or six trips his first month in the trade working with Kohse, he was more than pleased to find that his share amounted to over a thousand dollars. He figured he'd probably do even better with a boat of his own.

He started out by carving a model of a 35-foot-long craft with a narrow beam of 6 feet and then took it down to a local wooden boat builder, Tomotaro Yoneda, whose yard was set up three blocks above the Upper Harbour down on Chatham Street in Victoria. Schnarr then went down to Seattle to buy the most powerful motor he could find. He settled on a six-cylinder eighty-horsepower Marmon car engine with Bosch magneto and dual ignition. Once he opened her up out on the water, he was happy to learn *Moonbeam* was good for seventeen or eighteen knots, plenty fast enough to keep him well clear of the vessels the US Coast Guard had to work with at the time, which were only good for a meagre twelve knots.

Now he had the ability to deliver to any number of drop-off points anywhere from Port Angeles through Puget Sound and the San Juan Islands to as far east as Anacortes. Also, Schnarr pointed out that with this new and powerful boat, they were able to leave later in the afternoon, especially since there was little to worry about from the US Coast Guard during those early years of Prohibition. They would start out by running across from East Point on Saturna Island, and then head straight across the border and be halfway down the American shore by the time it got really good and dark. They also chose nights when the moon was full and bright, making for easier navigation. But as time went on, and the US Coast Guard beefed up its fleet, Schnarr began scheduling his runs when there was little or no moon at all or when the weather turned bad.

Schnarr said things went along pretty smoothly for a couple of years. He was able to work steadily throughout the fall, winter and spring, averaging four or five trips a month. In the summer, he got to take it easy. By the mid-1920s, the US Coast Guard had improved its fleet. Since Schnarr's runs were often some distance down Puget Sound, he began operating only during the winter months when the nights were long, dark and preferably with cloud cover close to the deck. Hopefully, if things went well delivering up their cargo, they would be back in Canadian waters by the time daylight broke.

Schnarr stressed that he had no shortage of customers. And "as time went on, more and more of the people in the trade got caught and spent some time in jail. But on the distribution end, there was always someone to take their place ... The demand for goods didn't let up at all."

But still, he wasn't exactly happy with his boat's Marmon engine, and especially after United States customs announced in the spring of 1924 that it was about to have ten 75-foot cutters constructed for use as patrol vessels throughout Puget Sound. Upon learning that the new vessels were to be powered by twin 250-horsepower Sterling engines capable of delivering eighteen knots, he decided it best to find himself a new, more powerful engine for *Moonbeam*. Schnarr always kept up with the latest boating magazines and happened to notice advertisements for a three-hundred-horsepower Fiat airplane engine. It sounded like just the ticket, so he went down to Seattle and bought one for five hundred dollars. *Moonbeam* became quite the hit in Victoria harbour once the engine was installed, since it was soon recognized as the fastest boat around. Schnarr recalled that there would often be fifty or sixty kids lined up at the wharf to watch him fire it up and go. He also remembered how amusing it was later on in life to be approached by a grown man who told him how exciting it was, as a kid, watching him take off from the harbour in *Moonbeam*.

Even though *Moonbeam* was a good, reliable boat, Schnarr decided to build an even bigger one in order to increase his payload from seventy-five to a hundred cases. It was good timing, since he lost *Moonbeam* around this time when his partners, Tom Colley and Billy Garrard, were intercepted by two US Coast Guard cutters while making a run into a beach near Anacortes. The cutters opened up with machine guns and hit the engine, which slowed it down. Fortunately, the two rum runners were able to drive *Moonbeam* up on the beach and make off into the woods before the cutters caught up to the launch. (On February 17, 1925, Victoria's *Daily Times* reported that *Moonbeam*, "a vessel of 4.92 tons" had been captured off Samish Island by the US Coast Guard's *Wasp No. 267* and was in the hands of authorities in Seattle.) While his partners were able to make it home to Canada, Schnarr never saw *Moonbeam* again. Looking back on the incident, Schnarr realized he had one other advantage that led to his personal success in running liquor: "It was probably my excellent night vision as much as anything else that kept me out of the hands of the Coast Guard for all the years I was in the rum trade." If he'd been aboard and at the wheel, they probably wouldn't have lost her.

Luckily for Schnarr, his new boat, the 48-foot, 5-inch *Miss Victoria*, which he was again having built by Tomotaro Yoneda, was near completion. Now he was keen to power the launch with two, instead of one, three-hundred-horsepower Fiat Aero engines good for some six hundred horsepower, but found himself short of cash. The Fiat engines, which had only put him back five hundred dollars for *Moonbeam*, were now worth three thousand dollars each. Schnarr attributed the steep rise in price to the increasing demand brought on by rum running. Since he ended up being shot at by the US Revenue Service cutters a number of times, he always ensured that a steel plate was built in behind the wheel as armoured protection. It didn't take long for him to start showing off the bullet damage to it after pulling off a run across the line. Although Schnarr designed his fast launches and paid to have them built for himself, he didn't register them under his own name as a form of legal mitigation in the event one of them was ever captured. In *Miss Victoria*'s case, the reputed owner was a "Wm. A. B. Garrard, of 25 Cook St., Victoria, BC, boat builder"; in other words, his old partner, Bill Garrard.

In order to secure funds so he could buy the engines for *Miss Victoria*, Schnarr went to Tacoma to inform one of his steadiest customers, the well-known Seattle rum runner Pete Marinoff, that he was short three thousand dollars. A little while later, Marinoff stopped in at Consolidated Exporters' Victoria office (where Schnarr

Following the end of Prohibition, many of the purpose-built launches were to put work in other trades. The retired *Miss Victoria II*, likely renamed *Black Hawk*, is seen tied up to the float at the bottom of the gangplank at Irvines Landing where she was being used as a water taxi running scalers out to Gambier Island. Her Allison airplane engine kept melting the porcelain out of the spark plugs, shooting it like bullets and making enough noise to wake the dead. Photo courtesy of Howard White.

had been loading his cargo) and, without a second thought, reached in his pocket, pulled out a roll and peeled off thirty hundred-dollar bills and handed it all to Schnarr.

Once the two Fiat engines were installed, Schnarr was happy to discover that his new launch was good for about thirty knots with a full load on. Still, he wasn't all that happy with *Miss Victoria*, since he'd had her built her with a full cabin in order to add a stove and a couple of bunks. Trouble was, it made loading and unloading more labour intensive but, worse still, the cabin was a little too high for his liking, which made the boat easier to spot on the water. He ran her for around a year and then decided he needed a better boat and had Mr. Yoneda build him *Miss Victoria II* in 1927. This fast launch was 36 feet, 6 inches long and initially powered by a six-cylinder Fiat Aero engine which was only good for three hundred horsepower, with her registered owner once again listed as

"Wm. A. B. Garrard." Apparently, Schnarr found his new runner wasn't quite fast enough, so he replaced the Fiat Aero engine with a twelve-cylinder Packard motor good for 450 horsepower. It would seem that in order to further cover his tracks, Schnarr was not only registering his fast, purpose-built launches under someone else's name but was also perhaps hiding the fact that he was relying on the superbly powerful Liberty aeroplane engines for motive power, recognized as the rum runners' engine of choice for jumpin' the line.

Schnarr was delighted to discover that even with a full load of 125 cases, *Miss Victoria II* was good for thirty knots. After his experience with the full cabin on his previous boat, he had *Miss Victoria II* built with an open cockpit at the stern. Now he could store cargo out in the open, which made for easier loading and unloading, and he didn't even bother to cover it most of the time. "Nearly all our hauling was in the dark, so I figured it didn't matter too much. Also, the cabin itself was very low and only two feet above the deck since we wanted as low a profile as possible. I had a hatch at the top of the pilothouse that I could open. I would stand on a box and my head and shoulders would stick out the top and I then could see all around." Still, Schnarr wasn't entirely satisfied with this boat either, so he sold her to the Reifel family.

It looked like the law had finally caught up to Schnarr after Victoria's *Daily Colonist* reported on February 19, 1928, "Sheriff Jack Pike, of Clallam County, and his deputies today surprised two men unloading fifty cases of liquor from a speedboat bearing the name '*Miss Victoria*,' at Deep Creek, twenty-five miles west of Port Angeles, and captured the boat and its cargo. The men escaped in a skiff and the craft was towed into Port Angeles." However the *Colonist* mentioned that their investigation revealed that the seized speedboat wasn't *Miss Victoria* after all, since she had been tied up down on the Victoria waterfront for days. Two days later, the *Colonist* went on to report that the Coast Guard patrol boat CG-272 left Port Angeles that day for Seattle with 586 bottles of assorted liquors aboard, taken from the "fast speed boat *M-1586* [formerly *Baby Bottleman*, an American rum runner] when she grounded on a reef off Tree Point."

It would seem that by this late in the game, Schnarr's boats were well known to the US Coast Guard. A week and half later, on March 1, the *Colonist* reported that Schnarr's *Miss Victoria II* was able to elude the US Coast Guard picket *CG-269* at Sunset Beach near Anacortes amid a volley of gunfire. "Two men and a woman on the shore likewise thwarted capture, but the coastguardmen seized fifty cases of liquor and three automobiles and took one prisoner" in the encounter.

It was around the time all this was going down that Schnarr placed his next order for a fast launch with Yoneda. According to her Victoria ship registry entry dated August 7, 1928, the *Kitnayakwa* measured 45 feet, 8 inches in length, with her owner listed this time as a Joshua Gorry, 36-3 Avenue, West Vancouver, BC, fish buyer. Twin four-hundred-horsepower Lee engines were installed, since Schnarr wanted to ensure she was capable of carrying a payload of two hundred cases and still be able cruise at over thirty knots with the throttle wide open. (*Kitnayakwa*'s engine power was provided by two high-speed Lee Motor Company gasoline engines capable of delivering 1,036 horsepower to her twin screws, which were good for up to forty knots.) It was a good thing that Schnarr's new boat was particularly fast, since the American Coast Guard was getting more aggressive in its efforts to curtail rum running.

Now, they would only fire one warning shot across the bow as an order to stop, and if that didn't work, open up with their guns in order to disable the boat and capture her crew. And with more resources available to fight the seaborne trade in liquor, by the late 1920s the US authorities were beginning to have some success guarding the approaches into Washington state. Also, with the US government keeping up the pressure on Ottawa, Schnarr said, "we all knew that eventually there would be restrictions on running liquor right out of Canadian ports." This meant he was going to have to head some distance offshore to pick up his liquor orders. Now he would have to run a hundred miles out Juan de Fuca Strait from Victoria to rendezvous somewhere like ten miles off Pachena Point light, up along the west coast of Vancouver Island. There, for example, he would come alongside boats such as the 80-foot, 6-inch halibut boat *Chief Skugaid*, for his next order. This inconvenient development resulted in Schnarr having another fast but more seaworthy craft built. (*Chief Skugaid*, working as a distributor boat, was the longest-serving West Coast rum running vessel, having worked in the trade from March 1923 through to December 1933. Today's owner, David Cobb, keeps her moored up the Fraser River near Fort Langley.)

With faster Coast Guard cutters now in the game, Schnarr settled on a 55-foot, 3-inch hull with two 860-horsepower Packard engines and named her *Revuocnav* (Vancouver spelled backwards): like *Kitnayakwa*, a name Schnarr figured the authorities wouldn't be able to decipher if they saw her racing by. The powerful shore boat was built in November 1931 by Bidwell Boat Works, established by Tom Nakamoto that same year in Coal Harbour. Her owner was listed in the Vancouver Ship Registry as Coast Island Fisheries, Limited of Vancouver. As fellow rum runner Fraser Miles explained, she was of a step hydroplane design, with engine exhausts under water through the step, for quiet running. Also, when Schnarr was carving the model, he designed the bow with a deeper V, "more like a destroyer, so that it could cut through a wave rather than ride on it." Along with the extra length and beam of his new boat, he was going to be able to handle the swells of the open Pacific and still make good speed. Soon after launching, *Revuocnav* was making runs loaded with 250 cases and travelling at a speed of over forty knots. Schnarr was happy to learn in these twilight years of Prohibition that the US Coast Guard were only using their fast cutters during the daytime since they were worried about running into driftwood at night. Schnarr didn't let that bother him, since he'd gained lots of experience during his years in the trade repairing damage from hitting one thing or another in the water. He used *Revuocnav* right up until Prohibition came to an end in early April 1933.

Looking back, Schnarr realized that Prohibition was especially good to him but said it didn't exactly make him rich. By the time it was finally over he still had ten thousand dollars in the bank, "a fair amount in 1933" particularly since the Depression was well entrenched by this time, with a Canadian dollar worth around eighteen dollars in today's money. He also pointed out that he never bought or sold liquor himself, only hauled it. Schnarr was able to build his family a nice home out by Ten Mile Point in Victoria. And this was during the years when he also found the money to invest some seventy-five thousand dollars to upgrade to faster, more efficient boats. "But as much as anything else—the money and excitement and all—I enjoyed the challenge of the whole game that was involved in the rumrunning: the challenge of not getting caught." He felt that he did reasonably well because he made his own luck, never took any unnecessary risks and made sure he always had the best boats and equipment available at the time. In the end, he said, what he was doing wasn't bad for the Canadian economy either. He figured that during his years of rum running, he made about four hundred runs, averaging around one hundred and fifty cases a trip, and freighted around sixty thousand cases of liquor across the line, so he credited himself with bringing four million dollars of revenue into Canada. "And that was no small sum in the thirties!"

While Schnarr was quick to see the advantages in designing his own purpose-built hulls for jumpin' the line, Consolidated Exporters in Vancouver only got into the act in a big way once Captain Charles Hudson took over as marine superintendent in 1928. Under its new, dynamic leadership, Consolidated realized that it was better to build their own shore boats locally, since the fishing and work boats that they were relying on in the early years of the trade were much too slow, especially if they had any hope of dodging and out-manoeuvring US cutters. Captain Hudson said it was with this realization that they decided to start designing their own, smaller craft, "imitation of the boats the Americans had [for running booze from Canada to the US]. I supervised the building of most of them … like the *Fleetwood* now a beautiful pleasure boat!"

With its fine yacht lines, that new addition to the fleet was originally launched as *Skeezix* in 1929 by Vancouver Shipyards for the Reifel family's Pacific and Foreign Navigation. (The rum runner was named after a character in *Gasoline Alley*, a popular cartoon that appeared in the newspapers of the day. Skeezix is a cowboy term for motherless calves and was given as a nickname to the abandoned baby in the cartoon.) According to her entry in the Mosquito Fleet Notes column in the August 1930 issue of *Harbour & Shipping* magazine, *Skeezix* was double planked with red cedar above the waterline and fir below for the outer skin and measured 56 feet, 6 inches in length. Incredibly, she was powered by three, not two, 450-horsepower Liberty aircraft engines that were reportedly capable of driving her at forty knots, which some claimed made her the fastest rum runner out on the water at the time. Liquor was typically loaded into boats like *Skeezix* from the Reifels' Pacific and Foreign Navigation warehouse at the mouth of the Fraser River and freighted down into Washington state and probably out to the mother ships sitting out off the entrance to Juan de Fuca Strait. In 1941, eight years following the repeal of Prohibition, the owner of the yacht renamed her *Fleetwood* and today she is under the care and protection of the Britannia Heritage Shipyard National Historic Site located at Steveston, BC, on the South Arm of the Fraser River.

In a two-part story that appeared in *Harbour & Shipping* in 1990 and featured the locally built, wood-hulled, 65-foot, 6-inch *Tapawinga*, Hugh Garling set the record straight. He was particularly concerned with what was being said "with a pretence of

Skeezix, with her fine yacht lines and fast hull, running under power off Rum Row, Ensenada, in the early 1930s. Fraser Miles collection.

supposed insider knowledge, that the fast boats built in Vancouver were good for three or four years, then they'd be ready for the 'bone yard.' Nothing could be further from the truth." On the contrary, he argued, they were built to take on a very demanding job and, by detailing all that went into the construction of *Tapawinga*, Garling provided a good study of the high standards that these locally built rum runners were built to. *Tapawinga* was launched in July 1932 from the W. R. Menchions & Company Shipyard in Coal Harbour. Power was supplied by three eight-hundred-horsepower Packard marine conversions—"an incredible total of 2,400 h.p.!"— courtesy of Murphy Marine Engine Company of Los Angeles. These particular engines were designed by Packard, the automobile manufacturer, and were successors to the Liberty L-12. To feed these three gas guzzlers, *Tapawinga*'s fuel tanks carried between 2,750 and 3,000 imperial gallons of fuel. In its reporting on the trial run of *Tapawinga* in the summer of 1932, *Harbour & Shipping* said that the shore boat was towed out the entrance of Juan de Fuca in order to save fuel for a long voyage south. "She has no cruising

Once Prohibition was repealed, Captain Charles Hudson bought two fast rum boats, one being *Tapawinga*. He had her down in California in 1935 and was hoping to sell her when she caught fire at sea. Captain Hudson went over the side and was in the water for five hours before being picked up by the Coast Guard. Since the boat was still salvageable, he sold her off in Los Angeles, where she was converted into a sportfishing pleasure boat. Emmie May Beal collection.

Diesel, and her 1,500 h.p. gasoline engines are reported burning up about 110 gallons of fuel an hour at full power, though she will undoubtedly be driven at the most economical speed possible for a long trip."

Garling went on to say that nearly sixty years later, a former Menchions Shipyard worker told him that he believed *Tapawinga*'s top speed was about fifty knots! Quality was the key word, he pointed out, and "no expense was spared to ensure the vessel was well and strongly built, able to carry heavy loads, at high speeds, in all sea conditions, riding out the worst storms because she could not seek shelter in at an American port." Only the best material went into the vessel's construction and ten different kinds of wood were used, all fastened with copper or bronze, with the exception of a few galvanized iron fastenings. "Her lines showed her to be a fast, V-bottom, hard chine vessel with a fine entry, a moderate flare and a minimum of sheer."

Tapawinga was probably the highest development of that class of wood-hulled launch, known as a shore boat. In his book, Fraser Miles said that what with her incredibly powerful gasoline engines, she was the real queen of the rum fleet, not the "shabby old ship" *Malahat*, as many writers claimed. Emmie May Stone, who was aboard *Malahat* with her new husband, Captain Stuart Stone, while lying at anchor at Rum Row off the Mexican coast in 1932, was enthralled watching *Tapawinga* come alongside to load. Once the boat's holds were filled, she cast loose and the throttles opened up as she headed off up coast under speed. Emmie May claimed *Tapawinga* was most certainly the fastest shore boat of them all. This suggests that the launch was most likely being used for making runs right into California beaches to deliver her cargo. While *Tapawinga* proved particularly successful and made it through her rum running years unscathed, she didn't survive for very long past the Prohibition years. Put to use as a sportfishing boat working out of San Diego, she burned at sea in May 1935.

A particularly fast shore boat, the 90-foot, 2-inch *Taiheiyo* was launched from the yard of Harbour Boat Builders in 1929. In January 1930, *Harbour & Shipping* magazine said she was built as a "cabin cruiser." While the Blue Book lists a Harry Milner of Vancouver as her owner upon launching in 1929, a Francis Carmichael was listed as the owner later in 1929 through to 1931. In 1932, *Taiheiyo* was registered to a Consolidated subsidiary, South Seas Traders of Vancouver.

Hugh Garling described her as "long and low, and looking somewhat sinister in her battleship grey. She had the entry and forepart of a commercial fishboat which turned off to a flat run aft like a planing speed boat ... She was powered by an Atlas Imperial diesel as the centre engine for cruising and two Liberty aircraft marine conversions for the wing engines, well over 1,000 hp! This gave her a speed of about 23 knots, enough to show her heels to most Coast Guard cutters." While *Taiheiyo* was relatively safe from the "cutter-itis" affliction (being shadowed by a Coast Guard cutter), still, as a precaution, a rheostat was connected to her stern light. Then if she was being shadowed at night by the US Coast Guard, it was turned up slowly as they quietly pulled away. Then finally, when the light was at its brightest, it was doused, the power poured on and the cruiser raced away into the darkness.

Still, it was another of the fast shore boats, which Captain Hudson was instrumental in having built in 1929 (the same year *Taiheiyo* was being built at Harbour Boat Builders), that had one of the most interesting logbooks of all of British Columbia's purpose-built runners. This was the "cruiser" *Kagome*, another wood vessel designed with fine yacht lines and an especially fast hull, measuring 68 feet, 4 inches in length.

Hudson said that an American representing South Seas Traders approached him to build the boat. But according to the company's official incorporation papers dated October 6, 1931, Charles Henry Hudson was listed as managing director, and the company's two shareholders as "Charles Henry Hudson and Margaret Margach (Housewife)." In 1931, *Kagome* went on the books as owned by General Navigation Company of Canada, of which Hudson was also listed as a principal shareholder on incorporation, along with John James Randall. This shows how the liquor export corporations blurred ownership to cover their tracks.

Johnny Schnarr recalled that *Kagome* was powered by two four-hundred-horsepower Liberty engines and a seventy-five-horsepower diesel engine as an auxiliary for idling or cruising. He also noted that Consolidated Exporters had big plans for the high-speed runner. "They had twenty thousand cases of liquor stored in a warehouse in Ensenada, Mexico, and they wanted a craft capable of hauling twenty-five hundred cases at a time up the coast of California." When an "expert" from California hired to install the engines couldn't figure out how to wire them, Captain Hudson contacted Schnarr and asked him if he'd come over from Vancouver Island and finish the job. He didn't have any problem getting the engines hooked up, but he did say it took some time, as a lot of it was finicky work. Once she was finished, Schnarr—who established a profitable relationship with the company as one of the biggest distributors of Consolidated Exporters' liquor into Washington state waters from the Pither & Leiser warehouse in Victoria—and a crew of eight were soon on their way out the Strait of Georgia, around the bottom of Vancouver Island and headed for the open Pacific and down the coast to Mexico. But out past Cape Flattery at the seaward entrance to Juan de Fuca Strait, where they opened up the engines, they were disappointed to discover they would only turn 1,800 rpm maximum instead of the 2,400 rpm they were supposedly capable of. Schnarr quickly deduced that her propellers were too small and their pitch was wrong, so they returned to port to replace them. After that, *Kagome* ran like a clock and they made for the open sea once again to arrive three or four days later at Ensenada.

Once the 2,500 cases were loaded aboard *Kagome* from an Ensenada warehouse, they pointed her bow to a position two hundred miles north of Los Angeles. When the American speedboat showed up forty miles off the coast and was alongside, the guy in charge managed to talk Managing Director Hudson into running the whole load to shore in one trip with *Kagome*. The risky venture was easy to manage, since it was only a matter of standing off the beach and loading one of *Kagome*'s dories while the other was being unloaded ashore. Since the job proved a success, Schnarr volunteered to make a second trip down to Ensenada. Trouble was, he suffered badly from seasickness out on the open ocean, so Captain Hudson arranged to have him sent home.

In her book, *Personality Ships of British Columbia*, Ruth Greene said that the cruiser was running her cargos right into shore at night "under care of a bribed policeman. One day the lawman with the greased palm fell ill, and his replacement lined up the captain and crew and put them behind bars." (Captain Arthur Lilly was in charge of the *Kagome* at the time. Lilly was a long-time rum running skipper who started out in 1922 on the *City of San Diego*.)

In his autobiography, Fraser Miles included the cruise report for the Coast Guard cutter *Morris* in an appendix that detailed the seizure of *Kagome* on December 30, 1932. "8:30 a.m., sighted *Kagome* which was found to be fourteen and half miles southwest of Cabrillo Lighthouse ... Chief boatswain's mate Marion J. Stokes boarded the Kagome at 10:40 a.m ... The master of the Kagome was unable to produce a manifest or other required papers [which were usually well hidden or chucked overboard if the Coast Guard was closing in] ... 6.55 p.m. [Coast Guard cutter] *Shoshone* got underway with *Kagome* in tow ... December 31, *Morris* following *Shoshone* and *Kagome* to San Francisco ... 12:30 p.m. Arrived off Pier #26." It is not known what type or quantity of liquor was aboard when the *Kagome* was seized.

During the trial, when the defence argued that *Kagome* was more than one hour's steaming time out at sea when caught, she was taken out for a trial run. With her engines in bad shape and loaded down with five hundred cases of Scotch, the point was made; she was indeed out in international waters and *Kagome* and crew were set free and returned to Vancouver. While the defence won their case, Fraser Miles said he'd learned that *Kagome* was actually only twelve nautical miles from shore when captured in broad daylight. It remains a puzzle as to why the *Kagome* was so near shore during the day.

In his column "Times Past" in the February 17, 2002, issue of San Luis Obispo's *The Tribune*, Dan Kreiger explained how Canadian liquor, sitting aboard large ships anchored off the small ports of Pismo Beach, Avila, Morro Bay and Cayucos, was run into shore and handled. Canvas-covered trucks waited on the edge of the beach to load cases of alcoholic beverages, which they would then transport to friendly dairy farms to "hole up" until the next evening, when they took off for Los Angeles. Kreiger said that one local family, the Spooners, was unfortunate enough to get caught with whisky, which was being warehoused at their ranch down at Spooner's Cove.

At the same time, making fools of the "Revenuers" was seen as good sport. One of the favourites was the old "rocks in the truck" trick. Krieger heard an amusing tale of this prank at a County Historical Society meeting in Morro Bay in 1984. One night when the flashing lights of a mother ship far out at sea could be seen off Estero Bay just north of Morro Bay, the locals quickly caught on that the Revenuers were watching through their binoculars from the hills above and the party line system of telephones connecting all the local dairies was soon abuzz. A plan was quickly hatched and dozens of farmers were soon headed down to the beach in their Ford Model As and trucks. Then fishing boats started coming ashore and a full hour was taken unloading heavy crates into the waiting vehicles. Once loaded, the caravan headed off up to the main highway with the Revenuers in hot pursuit. Twenty minutes later, a new fleet of vessels arrived at the beach and this time the familiar trucks pulled up, loaded their payload and quickly made off for the dairy farms. Meanwhile, back out on the highway, the Revenuers were pulling over "innocent farmers" and checking their trunks and pickup beds only to uncover crates of rocks from Old Creek. And the farmers' explanation? Well, the rocks had been used for traction down to a sandy beach where they had been having a friendly clam bake!

Overall though, Kreiger noted, most of the local population continued to regard the Eighteenth Amendment, the Noble Experiment, as just "so much foolishness." What it meant locally was a little extra revenue throughout the 1920s, a bad decade for the agriculture industry, and when the US Senate passed a resolution submitting an amendment for repeal of Prohibition on February 17, 1933, there wasn't all that much rejoicing.

KAGOME DELIVERS TO THE BEACH

In 1966, Frank J. Hyman published a thirty-five-page booklet titled *The Story of Rum Running* as part of his Historic Writings series, which he described as, "A recording of the facts and descriptions concerning the area on the Mendocino Coast in and around Fort Bragg, California, from the memories of one who was active in pioneer operations of the lumber and fishing industries and the development of the Fort Bragg area." In it, Hyman provided a detailed account of how the *Kagome*, now running as a shore boat, made outright illegal runs into the beaches just north of San Francisco in the fall of 1932. (In his third-person recounting of the tale, one wonders how Hyman came to be so knowledgeable about the specifics required in organizing such a risky and outright illegal undertaking and, most of all, how he was able to describe the events and individuals involved in such great detail.) As such, the story proves a fascinating insight into how rum running and bootlegging operations were managed once inside American waters. While the liquor export trade appeared to be run and operated in a supposedly open and civilized manner in Canada, across the line in the US, where it was outright illegal, it was an entirely different story.

In his opening paragraph, Hyman introduces us to the lead characters: Sal, "a dark, heavy-set North Beach Italian who ran a 'speak-easy' or bootleg joint down on Columbus Avenue," who informs Tony, "a small middle-aged smartly dressed man ... who had been a prosperous businessman, later the big shot of Pacific Coast Rum Running" on how the Canadians are going to deal with Coast Guard and how their cargo order from the mother ship would be delivered up to shore contacts. As the story progresses and the bootleg gang set up a delivery, we learn the Canadian boat they've made the arrangements with is none other than *Kagome*. (This is most likely one of the three cruiser-type boats that Sal says a Canadian "big shot" was having built.)

Fraser Miles declared that *Kagome* was soon recognized as the best known of the Canadian shore boats, 68 feet long and powered by three 450 horsepower gas engines. Hugh Garling claimed that she was powered by two Liberty and one diesel engines. Fraser Miles collection.

Tony leaves for Vancouver to charter a boat and contract for ten thousand cases to be delivered. Meanwhile, Sal heads over to the coast to check out delivery spots on the outside coast closer to San Francisco. Once they decide on a location, they have to wire the mother ship and have her change her current position sitting off Monterey to a point west of Point Reyes. They have made arrangements with a wireless station to broadcast these messages twice daily, then the runners will know what to do at all times and, with a few hours' notice, can hit any spot. Tony informs Sal that they have to furnish a "shore pilot" for the Canadians, who are unfamiliar with that part of the coast, so they hire an old friend in Seattle.

They hire Captain Larson "a rough Swede that used to smuggle Chinamen when he was Captain on the Oil Tanker by exchanging sailors in San Francisco as per instructions from the Big Shot of Chinatown." Tony promises to pay Larson's $150-per-month salary and a bonus of fifty cents per case for all cases landed. Larsen says that this arrangement is okay with him and that "he would show the 'limies' how to find the spots in the dark." (At the time, most of the crew aboard the Canadian ships were most likely of English and Scottish background and, as such, were referred to as limeys. Limey or "lime juicer" was a term originally given to those who served in the Royal Navy in recognition of the fact that lime juice was added to their daily ration of watered-down rum.) Also, Larson recommends that Sal better buy a couple of speedy trucks, large enough to carry two hundred cases each, along with a couple of used Cadillacs to distribute the liquor around the city. This means they require another truck driver, along with "a couple of triggermen who know why we're furnishing them with 'gats.'" ("Gat" was a Prohibition-era slang term for a gun and often specifically for a Thompson submachine gun, a.k.a. a "tommy gun.") Among the crew they pick up are a truck driver named Lardo, "a young, dark, fat, pig-eyed Italian from the slums of North Beach"; Snipe, a fellow gunman, "a small, sharp, dark-eyed youth of Italian descent raised with North Beach hoodlums"; and a couple of real gunmen, the infamous Jimmie "Baby Face" Nelson and John Paul Chase.

Several days later, Tony rushes into Sal's with a wire in hand: "'The boat is leaving – stop – get shore pilot out to mother ship at once – stop – located sixty miles west of Point Reyes – stop – otherwise advise.' This is the long-awaited day. To them it means action, money and fulfillment of their ambitions to be again the Big Shots in the Rum-Running racket." Everything goes according to plan and in less than two hours, five hundred cases are unloaded from the dory running between the beach and *Kagome*. Over the next month, three more loads are landed, but by then the spot is getting hot, since they've been working the location too long.

One night, *Kagome* arrives in at Salt Point and is about to unload when the "Federals" make their appearance on the road above the beach. As they approach *Kagome* and bullets start hitting the deck, the Canadians get too excited and have trouble getting the engine started, which gives the dory time to run alongside just as the first engine fires up. Once the dory is aboard, they head out to sea under full power. Meanwhile back on land, the Federals make a hurried survey and discover that they'd be able to capture four men, two trucks and four cars, along with five hundred cases of Scotch. After more help arrives, the liquor is broken up on the beach while the trucks, cars and prisoners are taken into Santa Rosa where the prisoners are all put behind bars. After rolling around off the coast for three days, the Canadian shore boat finally receives a message: "Hit the Noyo at 12 tonight, will use same signals."

By Christmas 1932, the gang is concerned that the Noyo River drop might be getting too hot, so the operation is shifted up to Shelter Cove in lower Humboldt County, where five more loads are dropped. Then two days before the new year, *Kagome* is captured.

Things didn't go well for the San Francisco gang following the repeal of the Volstead Act. Frank J. Hyman concluded his story with a report on what became of them all two years later. Tony, "the dressy leader, who made millions in the game" was broke and playing the "Ponies"; Sal went back to running his nightclub; Lardo the squealer, Baby Face Nelson and John Chase at times ganged up with Dillinger and Pretty Boy Floyd, pulling off jobs in widely scattered sections of the country, until the Federals declared open season on them, killing Dillinger, Pretty Boy Floyd and Baby Face Nelson on sight. Chase was convicted for the murder of FBI agent Samuel Cowley and sentenced to life. Lardo knew that they would give him the works when they caught up with him, so he gave himself up and turned state's evidence, hoping by stooling to save his neck. He finally ended up behind bars where he was beaten up badly. And the final sentence in Hyman's rum running tale? "There is no peace for a gangster stoolie until he is returned to his maker." Ending on this note, Hyman certainly comes across as well apprised of how rum running and bootlegging operated. And since he seemed so familiar with the gangster element involved, one can't help but wonder if perhaps he was much closer to the action than he let on.

On the other hand, the Canadians involved in the adventure returned to British Columbia to follow more mundane careers and less high-risk ventures. As for *Kagome*, perhaps one of the more successful

Canadian-owned shore boats, she could still be seen in British Columbia waters well into the 1990s. Captain Hudson recalled that sometime after the Prohibition years, "she finally ended up at a big sugar refinery [the Rogers family's BC Sugar refinery] as a yacht; that's how smart she was!" But he added that he had the pleasure of running her "for many, many years." The Rogers family obtained the fast launch in 1935, and then in 1947 a James A. Moody of Vancouver bought the vessel and renamed her *Salt Mist*. After passing through the hands of various owners over the years, she was at last word apparently still afloat as the yacht *Absolute* in Puget Sound.

⌃

Chapter Twelve

SHABBY OLD QUEEN OF RUM ROW AND THE HALCYON YEARS, ENSENADA (1928–33)

If you have a power boat and wish to visit Rum Row, steer for a point about twenty miles down the coast from Ensenada, and five miles south-southwest of Point Santo Tomas. You will find many vessels there, and they will supply you with whatever liquor your heart desires, so long as you agree formally not to take it into the United States. On one of the vessels there is a quantity of Mercier champagne, 1919, American brut, in pints and very cheap. Moreover, this champagne has been aging at sea for five years! If you prefer a dry Roederer 1921, a Chateau Margaux 1923, or anything else, from Liebfraumilch to vodka, or from Chinese wine to Bernkasteler Doktor, it will be delivered to you nicely packed in burlap bags. And if your requirements are for whiskey, the cheaper bourbons and ryes will be delivered on to your boat at from $12 to $20 a case …

The men of Rum Row are principally tugboat men and fishermen who formerly worked in the inland waters about Vancouver. They come south for a year or more at a time, some of them work on the base ships anchored off the coast, and others to handle the fast ex-submarine chasers that carry the liquor to points off the United States where small speedboats come out and unload them. Most of these jobs are entirely within the law—in fact, all of them are except those on the few boats which "run in"—that is, go into American waters to unload. This is very dangerous, for the speedboat operators ashore resent having their lucrative jobs taken away. They revenge themselves by helping the revenue men capture such boats.

—ROBERT DEAN FRISBIE, "RUM ROW: WESTERN"

After his misunderstandings with the US Coast Guard regarding the mother ship *L'Aquila* were finally sorted out, Captain Stuart S. Stone returned to Vancouver in late November 1928, where he experienced a short break ashore from active sea duty. As Jim Stone noted in his biography of his father, no one was surprised when, in March 1929, Consolidated Exporters offered his dad the most important position in their fleet: command of *Malahat*, owned by Archie MacGillis's Canadian-Mexican Shipping Company. Stone took charge of the long-serving mother ship and was to remain responsible for her until 1933.

Built as a lumber schooner, *Malahat* was launched from a Victoria shipyard as a deep-water lumber schooner in the midst of World War I. Ten days after Britain and Germany went to war in 1914, export markets were cut off and all British shipping along the coast from Prince Rupert to Panama was paralyzed. For the coastal sawmills that had come to depend on the offshore trade, the world crisis was disastrous. And as the war progressed, the shortage of shipping was exacerbated by the government needs for vessels to transport troops and munitions. These requirements, along with losses to German submarines, drove freight rates to all-time highs. Lumber was stacked up in mill yards and the big export mills shut down, creating unemployment in BC, while just across the border in Puget Sound, mills were working steadily to fill orders for both local and California markets. The key to their success was that over the previous seventy-five years they had built up an unusual fleet of lumber schooners for both the coastwise trade and trans-Pacific markets. American mill owners were able to rely on over some three hundred schooners registered to Pacific coast shipowners. Shipyards in Puget Sound had been building vessels for years for small as well as the big operators like Pope & Talbot and the Port Blakely Mill, which co-owned and operated fleets of the schooners.

James Oscar Cameron and Donald Officer Cameron were expatriate Americans who were part owners of an export sawmill at Genoa Bay on southern Vancouver Island. The Cameron brothers were lawyers from New Mexico and Texas respectively, who had no prior experience in shipbuilding, let alone lumbering, when they arrived in British Columbia in 1907. However, they were at the forefront of a group of lumbermen who realized that if they were to survive, then the BC lumber industry needed its own fleet of carriers. They believed that it was possible to build these vessels in BC by drawing on the wealth of talent and expertise from just over the border. A December 1915 article in a Vancouver trade magazine by master mariner Captain H. W. Copp advocated the building of such boats. This sparked a meeting of the Manufacturer's Association of BC in which Captain Copp suggested that it would be entirely possible to construct suitable vessels locally from BC fir at a cost of sixty thousand dollars per vessel and the provincial government provided financial support for the building of a fleet of five-masted schooners to take advantage of the trans-Pacific trade in lumber, particularly to Australia.

Malahat was one of six wood auxiliary schooners constructed in 1917 by the Cameron brothers' Cameron-Genoa Mills Shipbuilders Limited, which, realizing the potential, established their shipyard in Victoria's Upper Harbour in 1916. These deep-water motor sailers were named after the first of the class to be launched on Canada's west coast, the *Mabel Brown* from Wallace Shipyards over in North Vancouver. The Mabel Brown class schooners were designed by J. H. Price, the former manager of St. Helens Shipbuilding Company in St. Helens, Oregon, who was appointed president of Cameron-Genoa Mills later that same year.

According to Victoria's *Daily Times*, "They are five-masted auxiliary schooners, with length along the keel of 225 feet, length over all, 260 feet, beam 44 feet, and depth of hold 19 feet. They will be equipped with auxiliary power, using oil fuel Bolinder type of engines, which will develop 220 horsepower, giving the vessels a speed, under normal conditions, of seven knots under engine power alone. Each ship will require a crew of fifteen men." Between 1917 and 1921, twenty-four of the five-masters were built at four BC shipyards: Wallace Shipyards of North Vancouver, William Lyall Shipbuilding of North Vancouver, Cholberg Shipyard in Victoria and Cameron-Genoa Mills Shipbuilders. One of these schooners was the *Malahat*, which would later move on from lumber freighting to gain notoriety as a rum runner.

The Mabel Brown five-masters were "bald-headed" schooners—that is, they had no topsails. The main sails were hoisted from the deck with steam winches, which reduced the need for experienced seamen. With war raging in Europe, good sailors were hard to come by. Besides their fore, main, mizzen, jigger and spanker sails, the vessels were also rigged with a fore staysail and inner, outer and flying jibs. The wooden schooners were all fastened with hardwood trunnels (literally "treenails": hardwood nails used to secure timber, planks and so on) and constructed with hanging knees (curved, natural crooks in trees, sawn and used most often to strengthen the hull-deck joint) from the forests of Vancouver Island. The vessels were all rated at 1,500 gross tons with a carrying capacity of around 1,500,000 board feet of lumber. Their two four-cylinder semi-diesel auxiliary engines were made by J. & C. G. Bolinders Mekaniska Verkstad in Stockholm, Sweden, and rated at 192 net horsepower each. Each cylinder of these "crude oil" semi-diesel engines had a diameter of 16.5 inches and strokes of 20 inches. Hugh Garling, who sailed on *Malahat* after she became a mother ship to the rum running fleet, considered the Bolinders "primitive," as did the engine room crew who had to deal with them. He noted that a blowtorch was welded to each cylinder head to enable them to be started when cold and the two exhausts ran up the spanker mast nearly to the masthead.

Hugh Garling, who signed on *Malahat* at the age of nineteen on January 20, 1930, recalled his first impressions once he stepped aboard. "I was struck by the wide expanse of her deck. The unbroken length from fo'c'sle to the poop deck must have been not much under 200 feet. Under sail, in a seaway, you could look down the length of her deck and see an enormous flexing, showing the tensions in her hull. You could also see the normal hogging and sagging stresses clearly, as well as the twisting, especially in a quartering sea. In the fo'c'sle, or below, all about you was the dissonance of sounds her timbers made as they worked and resisted the flexing. Her masts too, would exert other strains as the ship rolled or pitched."

Overall though, this unusual fleet of twenty-four five-masted auxiliary schooners was somewhat of a commercial failure in the trade they were built for. When the Great War came to an end, returning ships flooded the market, freight rates dropped and large sailing vessels quickly became obsolete since they were unable to compete with the modern lumber-carrying steamer or motor ship. Their carrying capacity was too low and with their unreliable, underpowered engines most ended up getting into trouble in foul weather. While these West Coast–built wood motor sailers had short and rather disappointing careers overall, the most famous of the group, *Malahat*, proved to be the exception. She pursued an active career in the deep-water cargo trade and made seven Pacific voyages from her launching in August 11, 1917, through to July 1922. In early 1923 she was lying idle in Seattle, but not for long. She was about to embark on another, more lucrative line of employment once Archie MacGillis returned the vessel to Canadian waters.

On May 1923, Captain R. G. Lawson gathered up a crew and headed down to Seattle to bring *Malahat* home to Vancouver. According to the ship's official Agreement or Articles and Account of Crew, dated May 7, MacGillis was listed as supercargo for the short voyage. Lawson was apparently already working with MacGillis at the time, since the document notes that the last ship he served in was *Trucilla*, which was also owned by MacGillis.

On June 30, *Malahat*, this time with George Murray in command, was loaded and ready to depart on its first voyage deep-water rum running. She remained actively engaged in the trade right up until Prohibition was brought to an end ten years later. After the halibut boat *Chief Skugaid*, which was involved rum running for eleven years, *Malahat* was the longest serving and, in all probability, the most productive vessel on the west coast throughout the US Prohibition years.

The prominent role the five-masted schooner played in the liquor traffic was highlighted in a detailed article that ran in the 1931 annual of the Canadian Merchant Service Guild. As journalist L. V. Kelly explained, "Vancouver's rum-running fleet is modest but effective, and the flagship of this active flotilla, modestly and decidedly shy, has a one hundred percent batting average. This distinguished craft is the Vancouver schooner *Malahat* ... The *Malahat* is the veteran, she has seen them all come and go, and she still lies far off at sea, under observation of the Coast Guard cutters, frankly or surreptitiously, as occasion warrants or opportunity arises, supplying the requirements from the tenders from shore."

In the early days of rum running, *Malahat* was reported to have made two or three voyages a year down the coast from BC as a mother ship, but this was later reduced to one or two as moonshine liquor became more readily available throughout the United States. The retired lumber schooner was well suited for her new line of work, since she was capable of taking on sixty thousand cases of liquor. When she left Burrard Inlet and put to sea, her hatches were plugged full and decks packed with case upon case of the choicest brands of Scotch, rye and brandy. Fraser Miles, who joined the rum running fraternity as a teenager in 1931, claimed that when they loaded his boat, *Ryou II*, from *Malahat* the summer of 1933, she was carrying eighty-four thousand cases in the hold with another sixteen thousand cases stacked on deck. (*Ryou II*'s job was to deliver the liquor from the mother ship or from either of the bonded warehouses in Victoria and Vancouver to shore boats, like the *Kagome*,

The *Malahat* steams through First Narrows in Burrard Inlet, propelled by her two semi-diesel auxiliary engines. When many of these Mabel Brown–class schooners found themselves undergoing difficulties in foul weather, it was quickly discovered that the ships were underpowered and the engines unreliable. Vancouver Maritime Museum, LM2007.1000.4707.

which would load orders destined for American customers and run them into a quiet beach. *Ryou II* was another example of what was known as a distributor boat.) By this time she was sitting out on her anchor and being used strictly as a floating warehouse that was resupplied from other vessels sailing from the trans-shipment port of Tahiti. In all likelihood, she wouldn't have attempted to sail down the coast with that much heavy cargo aboard.

On May 21, 1924, *Malahat*'s master, George Murray, and crew once again signed the official Agreement or Articles and Account of Crew for another voyage, ostensibly from Vancouver to Mexico and return. (Articles were the contract that the crew signed whereby

The *Geraldine Wolvin* was another five-masted auxiliary schooner built in British Columbia during World War I. Here she sails past Point Grey on her maiden voyage bound for the open Pacific, her holds filled and decks stacked high with West Coast lumber. Vancouver Public Library historical photographs collection, 20272, photo by Dominion Photo Co.

they "agreed to conduct themselves in an orderly, faithful, honest, and sober manner, and all times diligent in their respective Duties, and to be obedient to the lawful commands of the said Master ... in consideration the said Master hereby agrees to pay to the said Crew as wages the sums against their names respectively expressed.") After the ship was back in Vancouver in mid-October, a new agreement was signed on November 15 for another voyage from Vancouver to Mexico and return. *Malahat* returned to Vancouver on August 19, 1925, and, as Captain Charles Hudson mused, whether she ever did sail as far south as Mexico for those three voyages is entirely doubtful.

As one of a small fleet of mother ships, *Malahat* sailed south to Rum Row wherever the fleet was sitting at the time. In the early years, it was situated well out in international waters off the

Columbia River bar or Farallon Islands, just outside San Francisco's Golden Gate. But once it got too hot off the US west coast in the mid-1920s, the whole operation moved to Mexican waters to sit about twenty miles down the coast from Ensenada and about eight miles sou'sou'west of Punta Santo Tomas. Here, a sandy shoal eight miles off the beach in a fifteen- to twenty-fathom patch of water served as an ideal anchoring spot for heavily laden vessels. (Mexico continued to observe the universal three–nautical mile limit for territorial waters, while the United States had arranged a treaty with Great Britain with respect to liquor where the limit imposed was extended to twelve nautical miles or one hour's steaming from shore.) Hugh Garling said, "On our windlass, it would show 60 fathoms of chain out, giving that long scope for secure anchorage. It was sandy bottom providing excellent holding ground and ideal for transferring cargos from one ship to another, or as a semi-permanent anchorage for a mother ship such as the *Malahat*."

While the sensational captures and the subsequent trials of the owners and crews of sister mother ships *Quadra*, *Pescawha* and *Coal Harbour* dominated the headlines between 1924 and 1926, *Malahat* seldom made the news and her years in the trade remained remarkably uneventful and free of high-seas drama. Still, one small incident did garner some attention in the news. On January 4, 1925, a headline on the Marine and Transportation page of *The Daily Colonist* reported, "Malahat Thought To Be Lost When Caught by Storm." Captain George W. Murray of Victoria was in command at the time, with twenty-three Victoria and Vancouver men as crew, when they were caught in a bad storm off the California coast. "The last that was heard of the Malahat's activities was a report from Santa Barbara that county authorities have seized 100 cases of liquor believed to have been landed from the Malahat a short time before she sank. Two San Francisco men, Adams and Alexander, were arrested as alleged custodians of the whiskey consignment. An armoured car and truck were seized by deputy sheriffs." The following day, the *Colonist* reported that it had received a statement from officials of the ship's owner, Canadian-Mexican Shipping Company, declaring that stories circulating from the south about the mother ship's loss were entirely erroneous.

Captain John Vosper took charge of *Malahat* in 1926 and the Agreement signed on July 17 that year notes that this time it was for a voyage from Vancouver to Central America and return. Vosper, who remained in command for three voyages, noted, "the *Malahat* was a lucky ship," and that "she was best under sail, her engines being too small." As it was, he knew the vessel well since he'd served as first mate on a lumber voyage to Australia under the command of George Murray back in the summer of 1920, though probably on another schooner, not the *Malahat*.

In his short biography of Captain John Vosper, Hugh Garling described him as a "rum runner and a sailorman par excellence. Someone said that if a naval officer was always a gentleman and sometimes a sailor, then an officer of the rum fleet was always a sailor and sometimes a gentleman. Captain Vosper was a sailor and a gentleman ... Although an ex–Royal Navy man, he was quite casual in dress, and he did not in the least have a taste for 'spit and polish.' He avoided dressing in uniform, even to the hat, and invariably wore a felt fedora in northern climes, and when southward going, would break out a straw hat and, as like as not, would pad around in bare feet." In an article on *Malahat* that ran in *The Vancouver Province* in March 1955, L. V. Kelly said that "her appearance was not impressive, nor was her speed, yet she plodded along her earnest way, never faltered ... never blundered. This was probably due to the acumen of her skipper, a small, dark man who did more thinking than talking."

Vosper was born in a village in Cornwall in 1891 and arrived as a two-year-old in the port city of Vancouver where, growing up, he became fascinated with the sea and shipping. At fifteen years of age, he signed on his first ship, *Empress of Canada*, a Canadian Pacific

Railway passenger liner in the Vancouver to Orient service, as an ordinary seaman. During World War I he served as a sub-lieutenant in the Royal Navy and, when that conflict came to an end, returned home to join the rum running fleet, where he made a few voyages in the steam-powered rum running mother ship, *Kuyakuzmt*. Vosper's next berth was the steamer *Prince Albert*, but he also commanded other rum ships: the three-masted schooner *Fanny Dutard*, the two-masted schooner *Lira de Agua* and the steel motorship *Lillehorn*. Still, *Malahat* was his favourite ship.

Recalling his years as master of *Malahat*, Vosper described in Ruth Greene's book *Personality Ships of British Columbia* how it all worked: "We changed position on the California coast regularly. If a cutter came out and began to trail us, we would just head out to sea and run her out of fuel, we being under sail. When they dropped us, we would come about and wire for a position to meet the buyer's boats. It was customary to keep a good lookout from the cross-trees. The lookout would identify every vessel sailing our way, whether it was a fellow rum runner or a Coast Guard cutter. Technically the Coast Guard had no jurisdiction over us on the high seas."

He also explained how cargo transfers were carried out. "While lying off the coast of California, if we were loading a small boat, and had partially loaded her when a cutter hove in sight, we would put the liquor back on the ship. We always loaded on the offside so that the mother ship would intervene between the small boat and the Coast Guard cutter. This would give a chance for the small boat to put some distance between herself and the cutter before she was discovered. When discovered, there would be a mad chase. If it was a fast boat being loaded, she could easily zip off and escape the cutter."

In regard to those who served under his command on *Malahat*, "the deceptively soft-spoken" captain said, "We had a very good crew when she was a rum runner, with few drunkards. I remember having to lock up one of them up until there was a boat returning home to Vancouver, and we put him aboard."

Malahat's first rum voyage under the command of Vosper ended in December 1926 with their next voyage from Vancouver to Tahiti on February 22, 1927. Papeete was to remain a port of call right up until Prohibition came to an end in 1933. Following the passage of the restrictive act that required that a vessel actually arrive at the destination as stated on its papers, the Bronfman family interests combined with Consolidated Exporters and the Reifel family in order to operate this new liquor shipping arrangement of delivering up and loading well out in the Pacific at Tahiti, but as Peter Newman noted in his book *Bronfman Dynasty*, the triumvirate didn't get around to setting up a formal sales agency to operate out of Tahiti until April 1933. He also said that the stocks listed in the agreement included a full range of Canadian and Scotch whiskies, gins, American bourbons, European champagnes and liquors worth $1,230,396.45 at distillery prices.

As Fraser Miles meticulously documented in his book, *Slow Boat on Rum Row*, once the load in Vancouver, unload at sea, return to Vancouver system was abandoned because of the restrictive act, rum runners were actually able to enjoy their best years ever between 1929 and 1933. Miles recorded a threefold increase in rum runner ships at sea, from eight to twenty-five, and more than a fourfold increase in months out at sea by the ships in the fleet, from forty-seven months for a total of eight ships in 1929 to one hundred ninety months for twenty-five ships in 1932. One of the biggest problems encountered while mother ships were still sitting off the American coast for weeks at a time was that they were required to restock their stores of food, water and fuel. Since these supplies were often brought out from shore aboard American boats, they often found themselves running short if a us Coast Guard cutter was out there on station keeping a close eye on all their activities.

Captain Vosper completed three rum voyages with *Malahat*. Her second one, from Vancouver to Tahiti, was completed on the ship's return to Vancouver on July 8, 1927. Then on September 26 that same year, *Malahat* departed Burrard Inlet on a voyage from Vancouver to Malden Island and returned to Vancouver in mid-June 1928. (Malden Island lies in the middle of the Pacific just south of the equator.)

In February of 1929, Consolidated Exporters took over ownership of *Malahat* from Archie MacGillis's Canadian-Mexican Shipping Company and the US Coast Guard's old nemesis, Captain Stuart S. Stone, stepped aboard as master of the infamous mother ship. As his son, Jim, pointed out, no one was surprised when Consolidated offered his father command, for which he was to receive six hundred dollars a month. Of course, he accepted, but under the condition that he also be the ship's wireless operator, since he suspected that *Federalship*'s never-ending problems with the US Coast Guard were not entirely due to bad luck. On March 29, *Malahat* departed Burrard Inlet "on a Voyage from Vancouver BC to Ensenada, Mexico, for a period not exceeding twelve months," as it was declared on the ship's articles. As noted, after the seizure of *Federalship* in 1927, a major policy change by Consolidated Exporters had Rum Row shifted away from the American coast south to sit anchored off the Mexican coast.

Hugh Garling recalled that it was hard finding a berth in a deep-water ship during the hungry Depression years of the early 1930s. When he walked aboard the steamer *Arwyco* down at Vancouver's Burrard Dry Dock in December 1930, where she was in for an overhaul, he spoke with a crewman who directed him to an address on Hamilton Street. There he was to go into an office and speak with a "Captain Hutton." Of course, this was none other than Captain Charles Hudson who, after Garling introduced himself, said, yes, they were in need of another crewman and told Garling to be down at Menchions's Wharf in Coal Harbour at eighty-thirty the next morning and to look for the fast freighter *Taiheiyo*. Here he was to report to Captain Butler and sail down the coast to sign on *Malahat* sitting off the coast of Mexico. It was a special day for Garling. "My world had turned bright! I had a berth, I would be serving in sail, getting in my sea time and receiving pay above the scale for deep-sea sailors ... No doubt the *Malahat* would be calling at exotic foreign ports with many adventures awaiting me! I let my imagination run wild."

Apparently the trip south to Mexico was uneventful and when *Taiheiyo* arrived at Rum Row, the "smuggler's headquarters," Garling had his first glimpse of *Malahat*, sighting her five tall masts before her hull came into view. "She lay at anchor pitching gently as her bows plunged into the troughs of successive seas. I took in her fine sheer, her clipper bow and the shark's tail [a real one] fastened to the end of her long bowsprit ... She looked the part too, oozing with romance and adventure since more booze had run in her scuppers than any other, and she had endured a record ten years on Rum Row without being apprehended by the United States Coast Guard." At 1,543 registered tons, *Malahat* was also the largest of the mother ships until the latter months of Prohibition when the steamer *Mogul*, listed at 1,828 registered tons, arrived in at Rum Row to relieve her in June 1933. Still, he was curious to see what sort of swashbuckling shipmates he was about to become entangled with. Instead, he was happy to discover that they were all much like deep-water sailors anywhere: agile, bronzed and well muscled.

When young Garling climbed aboard *Malahat* on January 20, 1931, she was on her second voyage under the command of Captain Stuart Stone. Her voyage from Vancouver to Ensenada was to last from July 1930 to August 1931. He described Ensenada as a great haven and escape for rum runners. (He was probably referring mostly to the crews of Canadian shore boats and American fireboats that got to go ashore when they were required to lay over for eight or ten days during a full-moon phase when normal operations couldn't be carried out along the California coast.) Rum runners were always welcome in Ensenada, as they were free spenders. Still, "when you signed two-year articles on a rum runner's mother ship, it was much the same as a voluntary two-year sentence of penal

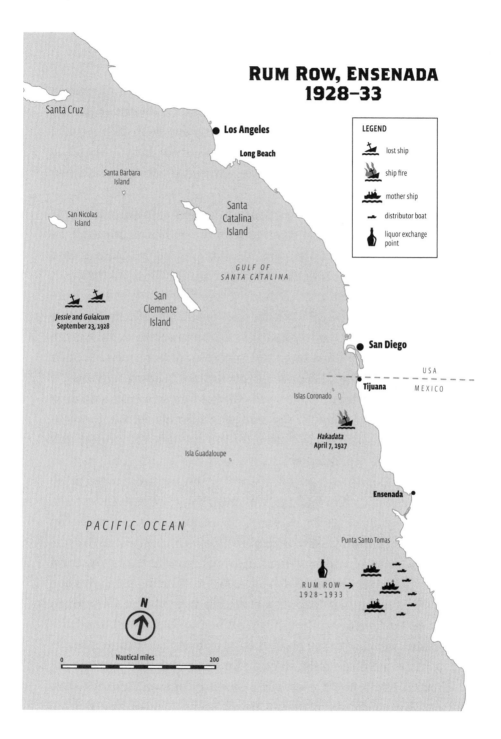

RUM ROW, ENSENADA 1928–33

LEGEND

- lost ship
- ship fire
- mother ship
- distributor boat
- liquor exchange point

Santa Cruz

● Los Angeles

Long Beach

Santa Barbara Island

San Nicolas Island

Santa Catalina Island

GULF OF SANTA CATALINA

San Clemente Island

Jessie and *Guiaicum*
September 23, 1928

● **San Diego**

USA

Tijuana   MEXICO

Islas Coronado

Hakadata
April 7, 1927

Isla Guadaloupe

Ensenada ●

Punta Santo Tomas

PACIFIC OCEAN

RUM ROW →
1928–1933

N

0   Nautical miles   200

Modified from map courtesy of Fraser Miles collection.

servitude. You agreed to obey the lawful command of the master which left you with little freedom of choice until the end of the voyage ... there was the endless day-after-day and week-after-week of waiting, drifting into month-after-month, for the day you would be homeward bound. On one voyage, I was at sea for an entire year without shore leave."

As Garling noted, life aboard a mother ship wasn't always filled with romance and adventure on the high seas. It entailed a lot of hard labour; in particular, dealing with all the cargo aboard required hours upon hours of hands-on work. Once all the cases were broken open and the bottles repackaged, the sacks were passed up or thrown along from one crewman to another in a line and spaced about ten feet apart. As Garling explained, "It was the fastest way of handling cargo and rarely did someone miss his catch. It was hard work and kept you in good shape." Then there was the problem of the long Pacific swells that prevented a large ship arriving with liquor in bond to run alongside a mother ship. In this case, the ship's boats would be used for the transfer. They could only carry around one hundred cases a load. The ship's boats were all rowed with long sweeps, but if a wind blew up, there was a small motorized launch to take them in tow. Then once alongside, there was the matter of getting the cargo up to the deck of the mother ship with a sea running down her length.

When this happened, the ship's boats would pick up the guest warp (a line for small boats to make fast to) hanging along the side of the hull and then load up a cargo net to be lifted aboard. Hugh Garling did mention that they'd all end up black and blue with bruises working in the ship's boats, as every sea running underneath would pick up the slack and then drop them down into the trough of the following sea, often knocking them off their feet. Then there was the hazard of the big, heavy steel hook used for lifting the cargo net aboard tossing around in the bottom of the boat and then a minute later high above their heads. Sometimes they'd end up with it dropping on top of their heads.

In 1932, Robert Dean Frisbie wrote an account of a voyage he made crewing on the so-called *Mariposa* under the command of a Captain McKlintock, from an unnamed French island somewhere in the South Sea to Rum Row off Mexico. The two-masted schooner was carrying a cargo of six thousand cases of bourbon, rye and Scotch whisky and "such unforgettable brands as Green River, Log Cabin, Old Crow, Black and White and Haig's Gold Label. It was all the *Mariposa* could carry. We had left our spare sails ashore so we could fill the lazaret with bagged quarts of Hill and Hill; our provisions had been piled on No. 2 hatch to make room in the pantry and the supercargo's cabin for cases of McCallum's Perfection; and even the main cabin. Had it filled to within a few feet of the deck carlings ... We felt that we were martyrs to a good cause."

In his tale, "Rum Row: Western," which appeared in the May 1932 issue of *American Mercury* magazine, Frisbie played it safe and assigned fictitious names to all the rum running vessels and those aboard, even though Prohibition was drawing to a close. *Mariposa* was likely the two-masted schooner *Aratapu*, a well-known ship that made regular supply runs from Tahiti to Rum Row. Frisbie noted that even though their voyage across the Pacific was pleasant enough, since it was summer with fresh trade winds, their imaginations often got the better of them. They were troubled by the ever-present fear of being set upon by hijackers, or the possibility that Mexican gunboats or US Revenue cutters might happen upon them and open fire. But when they finally arrived off the Mexican coast and made their way among the rum boats and US Coast Guard cutters, all anchored out in the same thirty-fathom patch, they were to discover otherwise.

They join a fleet of over a dozen ships, including "a five-masted schooner, a steamer, two beautifully equipped vessels that we took for rum-runners and a few smaller craft." Their liquor is consigned to the five-masted schooner, and in the morning Frisbee goes aboard her to meet "Captain Rockwell—a man responsible for running more than 1,000,000 cases of liquor into the States. He is probably the cleverest of the Pacific Coast rum smugglers, is wanted on numerous indictments, and is gall and wormwood to the Coast Guard. A few years on the right side of fifty, very alert and business-like, he is anything but the type of man one should expect to find in such an undertaking. He assured me that he seldom touched alcohol, and that only one drink a day was allowed the crew." It would seem that the five-master was none other than *Malahat* and Captain Rockwell, of course, Captain Stuart S. Stone, the persistent "gall and wormwood" to the US Coast Guard throughout the Prohibition years. (Stone would have been around forty-three years of age at the time.)

Captain Rockwell requests that the crew of *Mariposa* not give liquor to his crew. "I can't take a chance. This vessel is worse than a dump of dynamite, God!" Frisbie watches as Rockwell's eyebrows knit. "I have 60,000 cases of assorted liquor aboard—one hundred and sixty-three brands. Think what would happen if my crew once got started on it!" Still, while Rockwell points out that his vessel was nothing but a wholesale liquor warehouse and that you might read about "wild boozing voyages among the rum-runners: sawed-off shotguns, poker, fighting and drunkenness; but now-a-days this business is quite different. We are at sea for a year at a time, and that job is one long monotonous job of trans-shipping liquor. Of course, on the smaller boats [that run the booze into shore] it's different; they are cutterized now and then; every few months they put into a Mexican port for a big time, and sometimes they hit it up aboard their boats, too." Frisbie commented that overall, the rum running business was carried on smoothly and quietly while the Prohibition agents were out-manoeuvred. He also noted that the US Coast Guard's "fleet of pretty little cutters" ended up more of a benefit than injury to the rum running fleet. Since they set out to maintain constant vigilance over the fleet, they succeeded in keeping the riff-raff, the pirates and outright gangster element, at bay.

Continuing with his tale, Frisbie describes the next day, when a subchaser, the *Return Quickly*, comes alongside *Mariposa* to take on two thousand cases of Hill & Hill. At the same time, a Coast Guard steams over, stands off a few yards and holds her position until all the liquor is loaded. Once the transfer is underway, the captain of the cutter steps onto his bridge with a counting machine to tally the number of cases loaded. Meanwhile, aboard *Mariposa*, Frisbie and her captain start their own tally, and during the lulls in the work, turn to yarn with the cutter's skipper. Rockwell remarks to Frisbee that when a fire broke out on the rum runner *Toshiwara*, the Coast Guard was on hand and rushed to the rescue. Consequently, he calls over to the man on the bridge, "Thanks, Cap for helping us out." "That's all right," the enemy officer replies with a laugh. "That's what's we're here for; to help you fellows out and protect you from hi-jackers!"

The rum runner *Toshiwara* was most likely the two-masted *Hakadata*, which set it itself afire when it realized seizure was imminent after being spotted off Punto Santo Tomas by the cutter *Vaughan* in April 1927. (*Hakadata* was the old sealer owned by MacGillis's Arctic Fur Traders.) According to Frisbie, that night three Coast Guard men came aboard *Mariposa* to make one of their rare seizures: three bottles of King George IV Gold Label. "They drank it straight, out of the bottles, without chasers." The next day, after the remaining cargo was discharged aboard the five-master, and they'd taken on water, fuel and provisions, *Mariposa* set sail for Tahiti.

Malahat departed Burrard Inlet on her third run under the command of Captain Stone on December 12, 1931, on another voyage south where she was to lay for seven months as an anchored mother ship out on the Punta Santo Tomas grounds. During his third run to Mexican waters, Captain Stuart Stone brought along his new bride, Emmie May, née Binns.

The ever-popular and highly respected Captain Stuart Stone with his new wife, Emmie May. Emmie May Beal collection.

Hazel, Stuart Stone's sister, had met Emmie May at the British Columbia Telephone exchange where they both worked. Hazel brought her home to bunk with her in the Stone family's attic. Unbeknownst to Hazel, Stuart had already met Emmie at a party a year earlier. (Born in New York in 1909, Emmie and her family arrived on the West Coast in 1914, where her father, C. C. Binns, was already running a trading post in a small bay not far from Ucluelet. Captain Binns—who sailed around Cape Horn with shipmate Captain William L. "Whiskers" Thompson— had arrived on Vancouver Island in 1895.)

Following soon after *Malahat*'s return from her second voyage to Ensenada, Stone obtained a divorce from his first wife, who he had three children with. The divorce was finalized in October and Stone married Emmie May the following month aboard *Malahat* and then left for a short weekend honeymoon in Victoria. He then had to request permission from the directors of Consolidated Exporters to take his bride to sea with him. Now, with his brother Chet signed on as chief engineer, *Malahat*'s honeymoon voyage to Ensenada had three Stone family members aboard.

Emmie May told Dorothy Wrotnowski in a *Daily Colonist* story in March 1971 that when they finally put to sea loaded down with fifty thousand cases in her holds and five thousand tied down on deck and sailed past Brockton Point where hundreds of cars were gathered, they all dimmed their lights as they called out,

Malahat crewmembers and some of the ship's cargo. Hugh Garling collection.

"*Malahat* ... Farewell!" Soon the wireless network was abuzz with the news that the ever-popular Captain Stone was on his way to Mexican waters with his new bride. When they finally arrived at Rum Row, Punta Santo Tomas, they were delighted to see that all the men aboard the vessels on the Row at the time had shaved and were decked out in clean clothes in honour of the bride and groom. Emmie May also fondly recalled to John Lund in a 1994 article in *The Daily Colonist* how that first night she was bedazzled upon stepping out on deck to look out over a floating city of rum ships with all their anchor and cabin lights burning brightly against the dark shore. Stuart identified most of them: first there were the other mother ships, *Lillehorn*, *Nederiede* (Norwegian flag) and *Limey* (British); the smaller coast runners *Hickey*, *Hurry Home*,

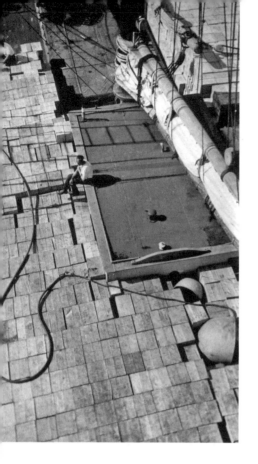

This bird's-eye view of the deck from up in *Malahat's* rigging shows just how loaded down the mother ships were sitting out on Rum Row. Emmie May Beal collection.

Taiheiyo and *Algie*; and two "strangers," a Tahitian schooner and another two-master from Panama.

Despite their wonderful reception, Emmie May said that the rum runner crews resented her presence at first. (Having a woman aboard ship was still considered bad luck that brought on bad weather and rough seas by many a sailor.) But after she took it upon herself to scale the schooner's rigging and tall masts in a rolling sea, without bothering to ask for permission from her husband, the ship master's and crew's attitude quickly changed to one of respect. Emmie May also recalled that crews "from all over the world" were soon stepping aboard to meet "Mrs. Capt." all dressed in their white shirts. Still, as Hugh Garling recalled in a personal interview in the late 1990s, the experience of having an attractive young lady aboard was somewhat trying for the crew, who were so far away from wives and girlfriends for months at a time. Often, while they were aloft working sails and rigging, they couldn't restrain themselves from looking down to sneak a peek at Emmie lying on the quarterdeck sunning herself.

First Mate James Donohue didn't speak directly to her for six months and would only communicate with her through other crewmen. As Jim Stone learned many years later, what probably annoyed Donohue was that his wife, Hazel, was an experienced wireless operator, and his request to have her come aboard *Malahat* had been turned down. (Donohue had married Hazel, Stuart's sister, in September that year with her brother as best man.) Donohue finally softened up once he saw that Emmie was fully capable of handling life at sea. She was so keen to fit in to shipboard life that she had a new hardwood floor put in the galley; helped paint the ship's tender, *Dixie*; and varnished *Patricia S*, the 22-foot speedboat that Stuart had bought for her and which they used for visiting neighbouring ships and going ashore. Donohue, who was overseeing renovation of the ship, even approved of Emmie's suggested colour scheme for *Malahat*: Irish green hull, red waterline, orange masts and white crossbars and top masts.

Malahat left Burrard Inlet on her final voyage south on December 9, 1932, only a month after Franklin Delano Roosevelt was elected president of the United States. At the Democratic National Convention in Chicago in the summer of 1932, the party adopted an election plank calling for the repeal of the Eighteenth Amendment, and when Roosevelt made his nomination acceptance speech he announced that from that day forward the Volstead Act was doomed. On February 17, 1933, just before Roosevelt's inauguration, the Senate passed a resolution to submit an amendment for repeal to state conventions. Meanwhile, out on Rum Row, *Nederiede*'s holds were empty so she pulled up anchor and left for Norway while *Malahat*, with thirty thousand cases aboard, *Lillehorn* and *Limey* were to remain on station until their holds were emptied.

The original plan was for *Malahat* to lay out on Rum Row until the summer of 1933 and then the steamer *Mogul* (originally launched as the collier *Caesar* in 1896), with Henry McKee Kennedy as her master, was to relieve her. (Hugh Garling, who signed on *Mogul* May 9, 1933, said the crew referred to Captain Kennedy as "His Majesty" after checking to make sure no officers were standing close by.) Captain Kennedy was to assume command of the *Malahat* and

With the need for another large mother ship, Captain Hudson convinced Consolidated Exporters to buy the old worn out steamer *Mogul*, which was launched as the *Caesar* in 1896. But ore-induced corrosion from carrying ore from Anyox, BC, to a smelter in Tacoma, Washington, had left her frames in very bad condition. So knowing she would never pass inspection, *Mogul* was put to sea where she was scheduled to stay for 12 months and then allowed to sink if it came to that. S.S. *Mogul* Jan. 29, 1932, Walter E. Frost photo. City of Vancouver Archives AM1506-S3-2: 447-2426.1

return her home while Captain Stone was to take charge of *Mogul* as the only mother ship left out on Rum Row. Although *Mogul* was in "god-awful condition" and her bottom pretty well shot with corrosion, Captain Hudson had recently bought her for Consolidated Exporters. She had been laying over at Ladysmith on Vancouver Island and she measured 310 feet in length and 1,828 registered tons. She was said to be the largest ship in service on the BC coast at the time. It was a timely purchase, since the two mother ships currently on station, *Malahat* and the steamer *L'Aquila* (previously *Federalship*), weren't large enough to take on the cargo if one or the other was to head home to Vancouver.

Mogul left Vancouver on May 30, 1933, to arrive in Rum Row later that June. No sooner there, *Malahat*'s crew quickly transferred

The *Malahat's* logbook entry for the passing of Captain Stuart Stone on July 5, 1933. City of Vancouver Archives.

their 737 cases of whisky over to the steamer. But just days away from taking on his new command, and having already sent over all of his and Emmie May's personal effects, Captain Stuart Stone suffered a severe attack of appendicitis. He refused offers to run him into San Diego, fearing that Uncle Sam hadn't forgiven him for his successful evasion of charges for his escapades off the California coast and, in particular, his nose-thumbing victory following the *Federalship* trial. When his appendix ruptured during the night of July 2, his brother and wife rushed him ashore into Ensenada. But since there was little in the way of medical help available in the small Mexican town, it was decided to fly him to Los Angeles. Unfortunately, it was already too late. The July 5, 1933, entry in *Malahat's* logbook reads: "S. S. Stone died this day of appendicitis. The First Mate J. V. Donohue has this day taken command [with] McGillivary Second Mate." Following the tragic loss, it was decided that Captain Kennedy was to remain with *Mogul* out on Rum Row, while the first mate, Stone's best friend James Donohue, took charge of *Malahat* and sailed her home.

On December 5, 1933, the fleet out on Rum Row heard breaking news over the California radio station KNX: It was all over and done with; the US Congress had finally repealed the Eighteenth Amendment, the Volstead Act.

It was while crewing on *Mogul* that Hugh Garling and shipmates finally got to step ashore at Ensenada. In later life, he described it as stepping into another world: "The people on the street, women, a tree, a blade of grass or a flower were all so unfamiliar." But once he got past the experience of having his feet on solid ground again, he quickly discovered that Ensenada could be best described as "Sin City" since it provided all that the wanton and debauched could ever desire. There was Cecilia's Place with the longest bar he'd ever seen on one side of the dance floor and all Cecilia's "dark-eyed beauties" on hand and eager to please. Outside in a courtyard were little huts in which the girls "befriended" tourists, tough old prospectors and sailors alike. By the early thirties, it became quite common and even fashionable for the Hollywood movie crowd to arrive and pick up their liquor directly from the rum ships. Well-known stars of the day such as Douglas Fairbanks Jr.; Georgie Jessel; Buster Keaton and his wife, Natalie Talmadge; Robert Montgomery; and John Barrymore aboard his yacht *Infanta* were all regular visitors, along with world heavyweight boxing champion Jack Dempsey, who opened the lovely Playa Hotel out along the beach. (Emmie May Stone recalled that when she and Stuart boarded Georgie Jessel's yacht *Kamin* with a case of champagne, his eyes lit up and he nearly knocked her over as he staggered past and went down on his knees to hug the champagne as if it were a long-lost lover.)

But it was riding off into the countryside that Garling got the most pleasure from, "where there were small farms, adobe houses and the ubiquitous donkey ... the real fabric of the country." One day he came upon a charming farmhouse set in the midst of an oasis of green. When he stopped to gaze and take in its beauty, he saw a very attractive girl standing on the verandah. He gave a friendly wave and she waved back, then walked down the path to greet him. Even though they didn't speak the same language, they communicated the best they could and Garling was soon convinced he'd finally met the love of his life. Later that day, when he wandered back down to Cecilia's and told his new friend Domingo, who occasionally arrived alongside *Mogul* for a case of booze, that he had a new girlfriend named Margarita and explained where the farm was, Domingo broke out laughing. "That farm belongs to Cecilia ... and Margarita is out there gettin' over a dose of clap!" Consequently, Garling didn't bother to keep his date at the cinema with Margarita that night after all. Besides Cecilia's Place, another Ensenada establishment that was frequented by the rum running fraternity was Hussong's Cantina which, as it happened, claimed to be the originator of the ever-popular drink, the margarita. Established in 1892, the cantina is still going strong today, but unfortunately the photograph of Emmie May Stone sitting next to *Malahat*'s life ring, which hung over the bar during the 1930s, is no longer there.

After spending over three weeks in Ensenada discharging her cargo into lighters alongside and taking on water, *Mogul*, the last mother ship left on Rum Row, departed on May 26, 1934, in ballast. After encountering strong winds on the voyage home, along with the handicap of her bottom being badly fouled with over twelve months' growth of barnacles and weeds, it was sixteen days before they arrived back to British Columbia waters. Once docked in North Vancouver, the old worn-out steamship was sold to Japanese owners who loaded her full of scrap metal. Japan, at the time, was buying up old steamers and sailing them home where both ship and cargo were thrown into furnaces and melted down.

Fraser Miles pointed out in his autobiography that of all the ships and boats that saw service rum running along the coast, it was *Malahat* that was crowned "Queen of the Rum Runners." He didn't exactly agree with that glowing testimonial, since he figured the glamorous title was all courtesy of writers who had never seen, and certainly never crewed on her. As far he was concerned, "she was a shabby old ship—the owners were as tight with paint as they were with sailors' pay. Crew members each received sixty-seven dollars and fifty cents a month and, in the end, had to seize the ship to get paid, a shabby kind of queen." Despite Miles's opinion, Jim Stone figured *Malahat* developed some of her reputation as "Queen of the Rum Fleet," simply for the fact that she was always being restocked with supplies. Besides her liquor cargo, the big schooner also served as a warehouse for everything the smaller vessels in the rum fleet might be in need of. As well as the thousands of gallons of fresh water *Malahat*'s condenser produced, there was fuel, food and medicine brought down the coast in vessels arriving in from the ports of Vancouver and Victoria, or brought out to the ship from Ensenada. In the end though, the five-masted auxiliary schooner proved to be as much of a success in the rum running fleet as she was as a deep-water lumber freighter. While a number of the West Coast mother ships, *Quadra*, *Pescawha*, *Coal Harbour* and *L'Aquila* (*Federalship*), ended up captured and seized by the US Coast Guard, *Malahat* sailed through the Prohibition years unscathed.

After the flagship of the fleet returned to British Columbia, she entered a new trade: log carrying. *Malahat* was lying at Burrard Dry Dock's wharf in North Vancouver with writ against her in 1934, when the Admiralty Court finally sold her off to William Clark Gibson for $2,500 in November that year. Based at Ahousat in Clayoquot Sound on the west coast of Vancouver Island, the Gibson family was involved in a number of ventures, primarily logging and lumber milling. Soon after purchase, the Gibsons transformed *Malahat* into what they claimed was the first self-propelled, self-loading and unloading log barge. They quickly set to work making alterations and fitting her out with steam donkeys for loading and unloading logs. As it happened, it was her old skipper, John Vosper, who took command upon completion, but after a couple of hair-raising crossings of Queen Charlotte Sound loaded down with old-growth timber, he decided it best to move on. After a few more close calls operating the under-powered log carrier in the treacherous and confined waters of British Columbia's coast, the Gibsons decided it was probably a lot safer to delegate *Malahat* to the towline.

The log-beaten barge finally met its end off the entrance to Barkley Sound, Vancouver Island, when she foundered under tow of the tug *Commodore* in March 1944. Tug and barge got caught in large swells off Cape Beale in a storm and the barge's load of spruce logs broke loose, pounding the bulwarks until the seams opened up. *Malahat* was consequently towed into Uchucklesit Inlet and abandoned. Canada's Department of Transport lost patience with the floating hazard to navigation and informed the owners in October 1945 that the derelict had to be removed. Old towboater Richard Simpson recalled that the Gibson brothers sent their tug *Joan G* to retrieve the hulk and tow it to the Powell River Company log pond over in Powell River, where she arrived on November 9, 1945, and her cargo of logs was unloaded.

The big pulp and paper company intended to use *Malahat* as part of the floating hulk breakwater protecting their log pond, but she was in poor condition, her hull having been badly beaten by logs being dropped in her. Consequently, she was put on the bottom off a rock breakwater at the southeast end of the mill's log pond and booming grounds in November 1946. Today the last of the old five-masted lumber schooner is a registered underwater heritage site and popular Powell River dive spot, where she still serves a useful role a century after her launch from the Cameron-Genoa Mills Shipbuilders yard in Victoria's Upper Harbour in 1917.

THE MYSTERIOUS LOSS
OF SS *CHASINA*

In her definitive 1969 book, *Personality Ships of British Columbia*, Ruth Greene devoted a chapter to the circumstances leading up to the mysterious disappearance of the steamer *Chasina* in September 1931. She begins by writing that SS *Chasina* left the port of Vancouver earlier that year on a mystery voyage, "loaded with food, gasoline, a new radio, and crewed by experienced cheery seamen," never to return.

Chasina's owners, Mid-Western Shippers Limited, with offices on Homer Street in Vancouver, purchased the rundown vessel in Vancouver harbour for $1,700 and then announced they were putting her to work in the Ensenada tomato trade. Greene pointed out that, as badly dilapidated as the steamer was, in her early years she was one of the most magnificent deep-sea yachts afloat. Launched as the *Santa Cecilia* in 1881 from the yard of J. Elder & Company in Glasgow, the 142-foot yacht cost her first owner, the Marquis of Anglesea, 210,000 pounds sterling to build. The elegant yacht was soon cruising the waters of Europe with crowned heads and British noblemen aboard. Sometime later, the dignified vessel passed into "bad hands" and became an outlaw on the seas with her name changed to *Selma*. In 1911, the All Red Line working out of Sechelt, BC, bought "the smart little ship" and after Union Steamship of British Columbia bought out the company in 1917, the steamer's name was changed to *Chasina*.

Once some Vancouver rum running interests, operating under the name of Alaska Pacific Shipping Company Limited, bought the vessel in 1925, "they painted the once proud beauty, a drab, battleship grey ... and she was now a lady of ill repute in the contraband game." She apparently wasn't all that successful, since she lay at British Columbia Marine Ways for two years until put up for auction for non-payment of berthage fees. After she was sold off to Mid-Western Shippers, they put her back running liquor but this time under the Newfoundland flag. (Newfoundland was an independent dominion at the time and didn't become Canada's tenth province until 1949.) Word soon got out that *Chasina* was steaming off to Macao across the Pearl River delta from Hong Kong. But what was she to load there was the question: opium, illegal Chinese immigrants or liquor? (In 1923, Canada's federal government passed the Chinese Immigration Act, which effectively banned most forms of Chinese immigration to Canada.)

In her research, Ruth Greene came upon a faded copy of the *Shanghai Times* dated June 18, 1931, along with legal documents and photos in some closed files of a Vancouver trust company. "Ship of Mystery, Romance, Pride of Nobleman Here on Strange Mission" read one of the headlines. "Battered and badly bent in places from a nine-day battle with the Pacific, a dirty, tiny, queer-looking craft, not unlike a common trawler, yet bearing signs here and there of having been built for something better than trawling, crawled into Whangpoo [Huangpu] yesterday, practically with its last gasp." When one of the hardened sailors was asked after the twenty-eight-day voyage across the Pacific what the reason was for their coming to Shanghai, he replied, "We understand that wines and spirits are cheap in this part of the world, compared to other parts, and feel that a shipload taken to profitable ports—Papeete, for instance, which is a free port—or the South Seas, or Mexico or some such place, might yield us good return." Although Captain S. Kitching was master of *Chasina* and her ten-man crew, it was none other than Captain William L. "Whiskers" Thompson, a shareholder with Archie MacGillis in Arctic Fur Traders Exchange, who represented the owners on board and was in complete charge of the vessel, although he only held a coastwise certificate. (Because of the MacGillis connection, one would presume that the two also had their hands in Mid-Western Shippers.)

According to Hong Kong's shipping master, *Chasina* left Macao for Ensenada on September 25, 1931, and after that she was never heard of again. The last cable Mid-Western Shippers received from the steamer was dispatched from Hong Kong on September 6 and read, "Due to arrive twenty-six days from date, radio schedule midnight Standard Pacific

Time." If the vessel was in distress after leaving the coast of China, no SOS message was ever received. What was particularly baffling was that the steamer had a newly installed radio, a shortwave Hartley circuit wireless receiver and sender, with a known range of between one and six thousand miles. During the presumption of death Supreme Court case held in December that year, it came out that *Chasina* had arranged to rendezvous with another vessel around one thousand miles off the American coast with a cargo of oil aboard for her. That vessel waited in vain for sixteen days at the arranged spot (47 degrees north latitude and 130 degrees west longitude) while continuing to call out her code over the wireless. John Holmes, the technician who installed the radio on *Chasina*, testified that if she had been anywhere near the arranged location, any wireless messages the ship sent out would have been strong enough for him to pick up in Vancouver. He never heard anything.

Whatever became of "the jaunty little pleasure yacht which once carried the crowned heads of Europe and ladies and gentlemen of society in the sunny seas of the Mediterranean" remains a mystery to this day. In concluding her tale, Ruth Greene suggested that it was entirely conceivable that she was followed out to sea and captured for the "thing" she carried aboard. Others speculated that perhaps the captain or owners had failed to pay the agreed amount to the dealers in opium or illegal Chinese immigrants or perhaps there was a fire or explosion aboard or the steamer had gone down in a typhoon. Regardless, no trace of SS *Chasina* was ever found, which all added to the speculation and folklore that followed upon her mysterious disappearance.

SUPPLYING A THIRSTY MARKET WITH THE CUP THAT CHEERS

So what did the whole rum running adventure out of Canada's West Coast ports amount to in the end? In its 1931 yearbook annual, the Canadian Merchant Service Guild included a glowing testimonial to the trade and commented on its significant contribution to the local economy: "Vancouver claims prominence in shipping on its grain and general traffic, but does not hold glaringly before the world its liquor traffic ... While tooting no horns and clanking no cymbals on the subject, the native Vancouverite will take no back seat from any rival rum-running centre. Its distilleries and breweries, its concentration and extracts, its fine corn products and rice distillations and fermentations continue to ooze out to go toward supplying any thirsty market with the cup that cheers ... Since rum-running became an industry and Vancouver took its place as a centre of material supply and executive administration, perhaps a million cases of whisky have passed from the Vancouver rum fleet to western American consumers, through the hands of United States boot-leggers, who buy their supplies from the Canadian liquor fleet ... Strange as it may seem, throughout the ten years that this traffic has been conducted from Vancouver there has been little disorder, but rare hi-jacking, and only one incident of major violence."

While the liquor export trade did indeed revive the Port of Vancouver and the West Coast economy, Stephen Schneider, author of *Iced: The Story of Organized Crime in Canada*, had a more measured take on the impact it had on the nation as a whole, both economically and politically. He pointed out that there was indeed a serious downside to the business and that the enormous profits that Prohibition made possible ended up creating what soon became history's single largest illegal market. If anything, once liquor was outlawed, it increased its desirability and in order to slacken the thirst of millions of Americans, Canada became a smuggler's paradise where bootlegging and rum running became nationwide industries. According to a US Coast Guard intelligence report, it was estimated that only 20 percent of the liquor produced in Canada was actually drunk in the country, while the other 80 percent found its way across the line into the United States.

In the early years, Canadian rum running interests were importing the bulk of their stock from overseas. According to the Dominion Bureau of Statistics, between 1924 and 1929 the value of liquor imports into Canada grew from $19,123,627 to $48,844,111. Still, Canadian producers were quick to realize that greater profits could be made producing liquor domestically and then exporting into the US market. The trade proved so economically rewarding that it led to the transformation of Canadian distilleries and breweries into some of the largest and most profitable liquor enterprises in the world. By 1927, there were eighty-three breweries and twenty-three distilleries operating in Canada, all licensed by the federal government. As Stephen Schneider pointed out, "The palms of government officials would be liberally greased, leading to an epidemic of corruption that to this day is unparalleled in Canadian history." Indeed, the corruption was so endemic that it led to the customs scandal of 1926, which brought down the Mackenzie King government.

Both the federal and provincial governments reaped substantial financial windfalls and benefited from a favourable balance of trade with the US by looking the other way throughout what would be best termed the Volstead prosperity years. Stephen Schneider's diligent research revealed that with their monopoly over liquor sales, annual revenues accruing to provincial governments from taxes collected on liquor sales and licencing fees for export warehouses increased from $3,837,000 to $22,755,000 from 1920 through to 1928. And, according to an article in *Maclean's* magazine in 1932, it was estimated that the provincial governments accrued a total of $152 million in liquor sales and taxes between 1922 and 1932, while the federal government reaped some $399 million.

Overall, Prohibition provided an excellent study with its curious juxtaposition of two nations sharing a border where a consumer product in particularly high demand is illegal on one side and legal on the other. In Canada, supplying liquor to the American market during Prohibition revived a flagging economy that was immersed in a postwar recession. Besides the economic benefits, it provided gainful employment for thousands, from all those working throughout the country's breweries and distilleries to those crewing on hundreds of boats and ships off the coasts of North America, in the Great Lakes as well as the Caribbean. But there was a major cost to all this economic growth. It helped to leaven and build organized crime into what remains today, where an often violent drug trade continues to thrive out on the streets of communities throughout Canada and the US. Hardly a week goes by in British Columbia without the local media running a story of another drive-by shooting in the Lower Mainland accompanied with photos of local gang members. Meanwhile farther south, some have suggested the incredibly profitable drug cartels transformed Mexico and Colombia into what might be best termed failed states.

As Stephen Schneider noted, Prohibition taught the criminal element an invaluable lesson: that there was more money to be made satisfying the vices of a receptive public, since it often meant there were no victims to complain to legal authorities. A 1929 article in *Canadian Forum* magazine explained how Prohibition gave a new rallying point and coherence to the criminal element. "There is a demand to be satisfied, and the legal sources have been stopped. The result is an illicit trade that has reached the basis of an established industry. But those engaged in it have to face competition, often accompanied by violent methods, and they cannot appeal to the protection of the law. So the trade has built up its protective system and in many cases its own legal system as well. It has acknowledged district heads, it has its own courts, attorneys, judges, it has its own armed forces; and the court of gangland is more binding on members, and its breach is more severely avenged, than the laws of civil society in which the gang exists."

To this day, the Noble Experiment is remembered as both a farce, and more importantly, a dismal failure. If anything, it was a time when people were drinking more alcohol and really enjoying it. Prohibition put the roar in the Roaring Twenties, those glamourous boozy years of rakish mobsters, flapper girls and crowded speakeasies. Even those who would otherwise never have considered deliberately breaking the law were simply ignoring the unpopular legislation in order to drink and have a good time. Still, the cost to American society was far more damaging than in Canada where it was legal, and by the late 1920s much of the American public were of the mind that the cure was worse than the disease. Once liquor crossed the line into the US, black market capitalism took over. Here the importing, marketing, sales and delivery were often left in the hands of the big-shot criminal element such as Chicago's infamous Al Capone and his charming bunch of gun-toting gangsters, with no rewards accruing to the nation's civic, state or federal governments. If anything, the costs of the belated and generally futile attempts to enforce the Noble Experiment were a drain on public treasuries. But once Franklin Delano Roosevelt and the Democrats came to power in the midst of the Great Depression, they were of the mind that after a flow of legal liquor was well underway, taxes on the commodity would swell government coffers. Indeed, the new source of funds helped to finance the New Deal, a series of programs initiated by the government to provide relief to the millions of unemployed and stimulate recovery of the American economy.

Meanwhile a similar situation is being played out today with the legalization of marijuana in a number of US states, where the number of overdoses resulting from the ingestion of nasty, toxic drugs, in particular opioids, consequently dropped soon after. Thankfully, Canada's federal government caught on and got with the program and legalized pot throughout the nation in 2018. As proved to be the case with liquor, as long as a popular consumer product remains illegal, it continues to be a major drag on the public purse. A 2002 report by Canada's Senate determined that the annual cost for law enforcement and the justice system was somewhere between three to five hundred million dollars, a futile attempt to curtail the use and sale of the recreational drug. While marijuana is a relatively safe and harmless intoxicant, perhaps it's time to re-examine all of our drug policies and laws, especially with the very serious opioid-fentanyl crisis underway, which has spread throughout North America.

Despite the fact that marijuana has been legalized, we are still a long way away from solving our drug problems. As long as there remain exceptional rewards to be reaped through the production and sale of other far more dangerous and addictive intoxicants, there will always be certain individuals who are more than eager to step forward to meet the demand. Of course, with hard drugs continuing to be illegal, the trade has evolved into a particularly violent one, much like Chicago throughout the Prohibition years.

Still, there is another reason prohibition only leads to greater costs to be borne by society stemming from drug use, and opium provides a good example. During the nineteenth century, Great Britain tried to stop the importation of the drug from China, which resulted in a trade war between the two countries after the transportation of opium was restricted. This only led to the production of heroin, simply because it was far easier to conceal, transport and thereby avoid detection once morphine was extracted from dried poppies and packaged up in much smaller bundles. And today we have fentanyl, which is a thousand times more potent than morphine, with grains that are minuscule and next to impossible to detect.

It is most unfortunate that we in North America never took a serious look back at what went down throughout the 1920s and early '30s and realized that Prohibition was a failed experiment and outright disaster. Still, there is one nation that has been proactive and realized that banning popular as well as addictive intoxicants simply doesn't work. Back in 2001, Portugal decriminalized the possession and use of small quantities of drugs—pot, cocaine, heroin, you name it—and chose to treat the problem as a public health issue, not a criminal one. And the result? The small nation's drug-induced death rate subsequently plummeted to five times lower than the European Union average. That being said, more governments should consider the fact that Prohibition in the United States in the last century did society a great favour by providing us with one of history's greatest lessons.

BACK HOME TO THE WEST COAST

WHAT BECAME OF THEM AFTER PROHIBITION?

Once it was a foregone conclusion that Prohibition was all but over, Canada's West Coast rum running fleet slowly began wending its way homeward throughout the summer of 1933. *Malahat* was back in Burrard Inlet that July and when *Ryou II* arrived on August 23 and the crew was paid off at the Consolidated Exporters office the next day, Fraser Miles said, "Everybody was in really clean clothes, and with complexions several shades lighter than they had been for months. How splendid to be clean again!" He continued by saying that *Kagome* left Rum Row, Mexico, in September, *Lillehorn* was back in November, *Tapawinga*, *Principio* and the distributor boat *Chief Skugaid* arrived in December while the smaller boats, *Shuchona*, *Algie*, *Audrey B*, *Hickey* and *Zip* all had returned by January 1934. *Mogul*, "a sort of holding tank without any distribution boats," was the last vessel to leave Rum Row and returned to Canadian waters in June that year.

And what became of all those who sailed with the rum fleet—its officers and crews who had been caught up in all the adventure, excitement and supposed money to be made? Many likely continued on in the marine industry and found work in more mundane and less exhilarating careers working aboard various deep-sea and coastal steamers or in British Columbia's towing and fishing industries. (Jim Stone suggested that the chief motivation for most of

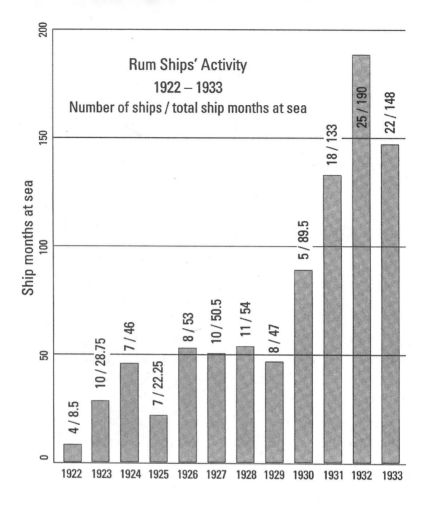

Rum Ships' Activity
1922 – 1933
Number of ships / total ship months at sea

Ship months at sea

4 / 8.5
10 / 28.75
7 / 46
7 / 22.25
8 / 53
10 / 50.5
11 / 54
8 / 47
5 / 89.5
18 / 133
25 / 190
22 / 148

1922 1923 1924 1925 1926 1927 1928 1929 1930 1931 1932 1933

This detailed graph of rum ship activity, listed year by year, reveals the major research Fraser Miles undertook following the end of Prohibition. Fraser Miles collection.

those who became involved in rum running was a postwar yearning for an exciting and relatively safe adventure in order to eradicate distressing memories of the World War I killing fields.) Most walked ashore after rum running with excellent credentials. If they hadn't been seasoned mariners before signing on with the export liquor trade, many had been able to accumulate an impressive amount of sea time, since Prohibition lasted some thirteen years. As Fraser Miles said, rum runners as a group made the Sphinx look like a

Fergie, George Butts and Sparky (left to right) homeward bound aboard the MV *Ryou II* in August 1933 after ten months at sea. Fraser Miles collection.

chatterbox, which "allowed the business to flourish in obscurity at the time, but left an impossible job for later historians trying to research Pacific Coast rum running, especially as the active participants didn't write anything down either." It's to the credit of the small handful of individuals who finally did put it down on paper or, at least, became more open to being interviewed in later life, that we have some insight into how they fared once it was over and done with.

While Fraser Miles scoffed at Captain Hudson's claim that fabulous amounts of money were made in the trade, the two thousand dollars he managed to save was put towards a good education.

Fraser Miles behind the oars on *Leviathan*. Fraser Miles collection.

Shortly after signing off *Ryou II*, he was on a bus headed to Tri-State College in Angola, Indiana. After finishing up high school there, he went on to earn a degree in engineering. Upon graduation, he stayed on in the States for a year or two and went to work for General Electric and Philips before heading home to Vancouver. Miles served in the Royal Canadian Air Force as a radar operator during World War II, and with the return of peace, got on with BC Electric (the forerunner of today's BC Hydro and Power Authority) and worked on the Peace River and Columbia River power projects back in the 1960s. By the later years of his career he'd moved up in BC Hydro to become assistant general manager (engineering). His son Jim said that while he was growing up in West Vancouver with the family, he really didn't know anything much about his dad's early years prior to Tri-State College. Then in 1977 his dad told him that he was going to write his memoirs. Since Jim only knew him as a guy leading a quiet, predicable life who headed off to work every day at BC Hydro with a shirt and tie on, he was rather startled to learn he was actively involved in high adventure rum running as a young man.

After his years jumpin' the line, Johnny Schnarr tied up his boat and headed off logging up near Ladysmith along the east coast of

Vancouver Island and sometime later made his way over to Nitinat Lake on the west coast. Since the lake was connected to the open Pacific by a fast-running tidal river, Schnarr was more than delighted to be next to salt water once again. He was soon back out there, this time in commercial fishing with his partner, George MacFarlane out of Bamfield, in his troller, *Eljo*, which he kept moored at his float house on Nitinat Lake (George was the father of John MacFarlane, creator of the Nauticapedia website). Sometime later, he headed over to Vancouver and bought the 34-foot *Margaret N*, which he worked right up until he retired in 1969.

Hugh "Red" Garling smiles through a life ring from the steamer *Mogul*. Hugh Garling collection.

Hugh Garling, who grew up in Vancouver, started out at the age of fourteen, working on coastwise vessels during the summer months, and then quit school two years later to pursue a sea-going career full time. As well as the rum runners *Malahat* and *Mogul*, Garling also crewed on such well-known coast vessels as *Rosebank* and *Baychimo*, and served as master on the Gulf of Georgia Towing Company's tugs. Upon stepping ashore, he went to work in the personnel office for the towboat companies. He served in the army during World War II, saw action in Belgium and Holland, and upon demobilization returned to the coast, where he became involved in the import/export business, then an insurance company, got involved with cement boats and finally in 1972 became a realtor. Still, his main interest was life on the water and he always remained an active member of the West Vancouver Yacht Club.

Another individual who remained actively engaged in the maritime industry was Captain John D. Vosper. After a rather unsettling experience trying to manage the Gibson brothers' self-powered (underpowered) log barge *Malahat* in December 1934 along the treacherous BC coast, Captain Vosper moved on to working on tugs. Then in the spring of 1940, he was appointed master of another five-masted schooner, *City of Alberni* (previously *Vigilant*), where he was given the nickname "Honest John D" by his crew. Vosper stayed with the lumber freighter for three-and-a-half years. But after she suffered a severe wracking when sailing down towards Cape Horn, he had to put back into Valparaiso, Chile, where the schooner was abandoned since there were no facilities there to repair a vessel of her size. This was a particularly regrettable development for Vosper since it had been his favourite vessel up until that time.

While Captain Stuart Stone didn't survive to share his story in later life, fortunately his son, Jim, wrote an excellent biography of his father, which provided a detailed account of those tumultuous years when his dad served as master of rum ships. Stuart's second wife, Emmie May, as well as Jim and Hazel Donohue, all returned to the west coast of Vancouver Island following the end of Prohibition. Emmie and her sister Phyllis found jobs at the Big Boy gold mine in Herbert Inlet just north of Tofino. Later, Emmie went on to work in the pilchard reduction plant in Kildonan, where she met her second husband, Charles Kerr, a seine boat fisherman. But she couldn't put up with all the drinking and partying that was part of working life along the West Coast back in those days, so she left to find work in Bellingham. There she met her last husband, Fred Beal, a logging contractor who was running a small sawmill. After writing her memoirs, Emmie passed away in June 1994.

In 1934, Hazel and Jim Donohue moved out to Long Beach, Vancouver Island, where they roughed it living in a tent for around a year and then found work as caretakers of a house overlooking the beach owned by an American millionaire named Lovekin. From there, they went on to build and operate Camp Maquinna with its cabins and camping ground out along the beach. (There were only two places to stay on Long Beach at the time; the other was Dick and Peg Whittington's Singing Sands resort, which was established in 1937.) Jim Stone said his aunt Hazel, who worked as a wireless operator for a year and a half in the wireless shore station operated by Captain Charles Hudson, was somewhat open to speaking with him when he and his bride were enjoying a three-month honeymoon in one of his aunt's cabins in 1947. While she did tell him about her late-night streetcar rides out to "Cappy" Hudson's Point Grey home, she refused to discuss coded messages she transmitted to Rum Row mother ships, attendant ships at sea and shore boats waiting to smuggle the liquor into American coves and ports.

Jim Stone said that Donohue was even more reticent than his wife. He did show him one logbook, a daily record of the weather and geographical position of the ship, but that was it. Later, he was to discover that all the rum runners kept two logs; the second, which was a detailed record of actual operations, was kept right at hand ready to be "deep sixed" if the US Coast Guard boarded the vessel. Also, as it happened, a mile and a half down the beach from Jim Stone and his wife's 1947 honeymoon cabin lived George Ford, captain of the mother ship *Quadra* when she was seized off the Farallones in October 1924. But Jim said that at no time did Ford ever speak about his years rum running, even though they rode together in a friend's car out to a logging camp to go to work during the last six weeks of his Long Beach honeymoon.

Another member of *Malahat*'s rum-running crew who retired out to Tofino only a few miles from Long Beach was Third Engineer Fred Sailes. And as for Jim Stone himself? He eventually went on to obtain a degree from the University of British Columbia. During the war years and through the early 1950s he served in the Royal Canadian Air Force and after retiring from the service became a professor of English at Waterloo College and the University of Waterloo from 1958 to 1986.

One individual who proved a particularly valuable source of information was, of course, Captain Charles H. Hudson, who quit rum running early. After one too many long days managing the fleet, he decided he had had enough and left his demanding job as marine superintendent of Consolidated Exporters Corporation Limited five months before Prohibition was brought to an end. Even though he had "a wonderful, wonderful time ... I just got fed up with them so instead moved down to Los Angeles to manage a distillery, which I soon got tired of and decided to return to Vancouver." Once back, it didn't take him long to open a boat brokerage firm on West Georgia Street, only a short distance from his erstwhile enterprises during the heyday of rum running. Hudson stayed with the venture up until World War II broke out and he joined up in the Royal Canadian Navy, where he was appointed a lieutenant (temp.) in the Royal Canadian Naval Reserve. He was assigned to HMCS *Naden* and in 1940 was given command of the patrol and examination ship HMCS *Malaspina*, an old Fisheries patrol vessel. He was appointed lieutenant commander (temp.) in 1943 and ended up at Pacific Command in Jericho (Vancouver) by the end of the war. Upon retiring with the rank of commander, while he wasn't to go to sea again, he did stay close to the water and returned to boat brokering until retiring in 1973.

Still, there was one member of the fraternity whom liquor apparently got the better of. This was Captain Arthur G. Lilly of the ill-fated schooner *Chris Moller*. Captain Hill Wilson, master mariner, marine pilot and author, recalled in a personal interview that when he was working as third mate aboard Western Canada Steamship Company's *Lake Tatla* on a voyage to Alexandria, Egypt, and into the Black Sea in 1947–48 with Arthur G. Lilly as her master, the ship was soon "in a shambles" what with the crew "a rough and tumble group." Consequently, when they stopped in San Francisco on their return to the West Coast, Lilly received a letter from head office, addressed to "Mr." Arthur G. Lilly, informing him he was being relieved of command. During the mid-1950s, when Hill Wilson was working with Straits Towing out of Vancouver, he said he often heard Lilly, who was working on fish packers at the time, over the marine radio.

What became of Archie MacGillis, the supposed founding father of the West Coast rum fraternity, following the repeal of the Volstead Act remains a mystery. He did sell off his legitimate business, the Vancouver Courtenay Transportation Company, to the Vancouver Barge Transportation Company in 1929, probably in order to devote more of his energy and resources to his rum running enterprises, but once that was all over and done with, he appears to have disappeared from the scene. The only evidence of what he became involved with after the fact is his registration of death dated April 4, 1961, held in the BC Archives, which lists his profession as retired prospector and miner, residence West Second Avenue, Vancouver.

Jim Kirkland, in responding to a letter to the editor which ran in the July 2017 issue of *Western Mariner* magazine asking the readership if anyone knew of what became of MacGillis, contacted the author to say he remembered him well. Kirkland said that back when he was a teenager in the mid-1940s his father was running the general store in Hope, BC, on the Fraser River, and MacGillis, who was prospecting for gold in the area with partner Sam Frisby, used to come in to pick up supplies. Since they were often short of cash, the two partners were running up a tab and not only that, convinced the store owner to grubstake some of their ventures. The two characters were living in a shack at the time and eventually did open up a silver mine producing low-grade ore up by Holy Cross Hill, not too far out of Hope. Jim Kirkland also suggested that the boys might have been up to some shenanigans on the Vancouver stock exchange promoting their mining prospects.

THE SURVIVORS: THE BOATS

W hat of the West Coast's rum fleet itself following Prohibition? Of course, most of the vessels went back to more mundane jobs working as towboats, fish packers, coastal freight and passenger vessels or halibut boats. But after serving a number of years of productive service, most were eventually broken up or abandoned on an upcoast beach if not on the bottom. Still, surprisingly there are a handful of survivors still around.

RYOU II

Once the services of the distributor boat *Ryou II* were no longer required, she was sold off and has been involved in BC's commercial fishing industry ever since. In 1934, Northern Chief Packers Limited of Vancouver converted the fish-packer-turned-rum runner into a table seiner. She worked all up and down the BC coast and passed through the hands of a number of well-known fishermen. Subsequent owners were Walter Carr and John Kosulandish, Vancouver (1935); Walter Carr and Alphonse Veljacic, Vancouver (1945); Mike Polanio and Gary Beldigara, Vancouver (1946); Mike Polanio, Vancouver (1950); Thomas Mosley, Prince Rupert, who probably did the conversion to a drum seiner (1967). Delmar Laughlin, Sointula, BC, bought *Ryou II* in 1972 and renamed her *Rick-son*. (When this author was working as a skiff-man on seiners out of Sointula in the seventies and eighties, he had no idea at the time that *Rick-son* was once a rum runner.) Laughlin sold her off to

Ryou II. Photo courtesy of Rick James.

David Lansdowne and his Waverly Fishing Company of Courtenay, BC, in 1994. It was Lansdowne who changed the boat's name back to *Ryou II*. While packing roe herring years ago, Lansdowne met a Japanese buyer who informed him, "*Ryou*—good name! King of the sea dragons!" But two other contacts of Japanese descent have said that it simply means "fishing." Since selling his salmon licence in 1996, Lansdowne has fished groundfish and hake quotas with

Ryou II and as of 2016, has been packing salmon, herring and sea cucumbers. (He also noted that the cucumber divers are most appreciative of the fact that the boat provides hot showers and meals.)

Ryou II was originally built in 1929 as a fish packer for Kitaro Nitta of Steveston, BC, at the Bidwell Boat Works Limited in Coal Harbour, Vancouver. While Mr. Nitta is listed as *Ryou II*'s original owner on the vessel's Ships Registry documents, Fraser Miles wrote that it was actually Henry Reifel, the prominent Vancouver distiller and brewer, who was the behind-the-scenes owner. Upon launching, the wood boat measured 60 feet in length, 17 feet, 4 inches in breadth and 7 feet, 8 inches in depth and was 37.88 registered tons. The original engine was a three-cylinder one hundred brake horsepower Atlas-Imperial diesel, which was good for eight-plus knots.

CHIEF SKUGAID

According to Fraser Miles, *Chief Skugaid* was the longest-serving West Coast rum runner, having worked in the trade from March 1923 through to December 1933. David Cobb of New Westminster, BC, currently keeps her moored up the Fraser River near Fort Langley. The original owner (1913–19) was Canadian Fish & Cold Storage Company of Prince Rupert, BC. She was originally built as a wood halibut schooner by Vancouver Shipyards, Limited in 1913 and measured 80 feet, 7 inches in length, 18 feet in breadth and 8 feet in depth and was 54.5 registered tons. Her single screw was driven by a four-cylinder Imperial gas engine producing 3.75 net horsepower. Along with her career as a halibut boat fishing the Bering Sea grounds, and her career as a rum runner, *Chief Skugaid* spent some thirty years as a fish packer. In 1926 she was owned by M. C. Smith of Vancouver. In 1929–31: General Navigation Company of Canada Limited, Vancouver. 1935: Henry J. Stump, Vancouver. 1937–64:

Chief Skugaid. Photo courtesy of Bob Olson.

Colonial Packers Limited, Vancouver. 1964–97: British Columbia Packers Limited, Richmond, BC. 2001–3: Jim Pattison Enterprises Limited, Vancouver.

SKEEZIX

Following the end of Prohibition, *Skeezix* was converted into the luxury yacht *Fleetwood*. Like a number of the purpose-built rum runners, *Skeezix*, with her fast, narrow hull, made for an attractive sea-going pleasure craft. She was built by Vancouver Shipyards, Limited in 1930 for Pacific and Foreign Navigation Company Limited. (The owners of Pacific and Foreign Navigation, the Reifel family, were operating a warehouse in Canoe Pass, Westham Island, at the mouth of the Fraser River at the time.) She was double planked with red cedar above the waterline and fir below. She measured 56 feet, 6 inches in length, 12 feet, 1 inch in breadth, and drew six feet of water and was 31.26 gross tons and 18.04 net. According to *Harbour & Shipping* magazine, the fast launch was powered by three

Skeezix. Photo courtesy of Rob Morris.

450-horsepower Liberty (aircraft) gas engines, rated at 450 horse-power each, while Fraser Miles claimed she was powered by one four-cylinder Atlas-Imperial diesel and two Liberty gas engines. Regardless, the engines were reportedly capable of driving her single screw up to forty knots, which some claimed made her the fastest rum runner out on the water at the time.

In 1932, ownership passed to Atlantic & Pacific Navigation Company (the Reifel interests) of Vancouver. In 1933, she was owned by Laredo Fisheries, Vancouver. In 1937, she was owned by Tom Wood, Vancouver. From 1942 through to 1966 she was owned by Jacob W. Cohen and from 1967 to 1972 by John B. Buchanan of Vancouver. From 1973 to 1975 she was in the hands of Norm Perron of Qualicum Beach, BC, and from 1976 to 2001 she was owned by Robert Turnbull of Edmonton. She was finally donated to the Britannia Heritage Shipyard in Steveston, BC, by the family of Robert Turnbull after he passed away. As of 2018, the vessel remains sitting in the National Historic Site's boatyard until the Britannia Heritage Shipyard Society determines what to do with her. They are hoping that perhaps the City of Richmond will take over ownership.

KAGOME

Like *Skeezix*, the shore boat *Kagome* was turned into a luxury yacht and was seen around BC waters well into the 1990s. Built for Captain Charles Henry Hudson in the yard of Harbour Boat Builders Limited in 1929, *Kagome* measured 68 feet, 4 inches in length, 13 feet, 1 inch in breadth and 6 feet, 7 inches draft and was 40.08 gross tons and 27.25 net. The runner was powered by two four-hundred-horsepower Liberty engines and a seventy-five-horsepower diesel engine as an auxiliary for idling or cruising. Finally, in 1931, *Kagome* was registered to General Navigation Company of Canada, whose principal shareholders upon incorporation on March 9, 1931, were Charles Henry Hudson and John James Randall. Charles Hudson said that after rum running came to an end, she ended up in the hands of the Rogers family's BC Sugar on Burrard Inlet and became their yacht, *Salt Mist*.

From 1935 to 1945, the yacht was registered to Phillip T. Rogers. In 1946 she was owned by Arnold G. King. In 1947 she was owned by James A. Moody. In 1948 she was owned by James Hoffer. All of these owners were based in Vancouver. Through 1950 to 1963, the vessel passed into government hands and was registered to the Minister of Lands and Forests, Victoria. In 1964 she was owned by William Reeves, North Vancouver; in 1965 by Ronald A. Roe, Vancouver; and in 1968 by Berton Enterprises Limited, Vancouver. In 1973, she was owned by Salt Mist Enterprises, Port Alberni, BC. In 1980 the owner was listed as Knut Leine, Tofino, BC, and in 1985, Allan E. Moore, Delta, BC. Then from 1988 up until 1993 she was owned by John Waterman, Victoria, and his company, Briarholme Holdings Limited. *Salt Mist* passed into American hands when Mr. Waterman sold her to Fred Jacobsen of Whidbey Island. Unfortunately, Fred Jacobsen died in a tragic accident a few years later and the family consequently sold the yacht off. John Waterman heard later that *Salt Mist*'s new owner renamed her *Absolute* and was keeping her moored in Lake Union, but another source says that she may have been towed out of Lake Union and back to Canada.

TEMISCOUTA

Temiscouta, a fish packer built in 1930 by Meteghan Shipbuilding Company, Meteghan, Nova Scotia, made only one Pacific Coast rum running voyage (under Captain J. D. A. Wood, with radio operator "Smoky" Hoodsmith) from November 1932 to March 1933, according to Fraser Miles. Once Prohibition was over, she sailed north to BC waters, where sometime later she was renamed *Riverside Y*. *Temiscouta* was originally built as a fishboat and was 60 feet, 10 inches in length with a beam of 13 feet, 4 inches and 7 feet, 4 inches in depth, with her registered tonnage listed as 24.5. A six-cylinder, one-hundred-horsepower Standard diesel engine drove her single prop. *Riverside Y* passed through a number of owners on the West Coast. The most recent of these was Lance Goertzen of Delta, BC, from 2003 to 2014. Recent reports have it that *Riverside Y* was a liveaboard high and dry in Squamish for a short time and then when a property developer bought the land above the beach where she was lying on the hard, the old packer was finally broken up.

YUKATRIVOL

The last purpose-built shore boat built in Burrard Inlet, *Yukatrivol*, was launched from the yard of Union Boat Works in Coal Harbour in 1933, and left on her first and only rum running voyage (under Captain C. R. Brewster) on April 21, 1933. Built for Arctic Fur Traders Exchange Limited (with Archie MacGillis and William L. Thompson as directors), *Yukatrivol* measured 59 feet, 10 inches in length, with a breadth of 12 feet, 8 inches and a depth of 5 feet, 10 inches. The fast launch was driven by three 450-horsepower Liberty engines.

Yukatrivol. Photo courtesy of Rob Morris.

From 1935 to 1951 she was owned by Higgs Gabriola Ferry Limited of Victoria, which renamed her *Chief Maquinna*. In 1951, ownership was transferred to William Y. Higgs of Nanaimo, BC, and she became the *Marine Explorer*. From 1956 through 1958 she was owned by North American Towing & Salvage Company of Nanaimo, and then from 1958 to 1975 by the Minister of Lands, Forests and Water Resources, in Victoria, and was renamed *Forest Cruiser*. From 1976 to 1983 she was owned by the RCMP and from 1983 through 2001 by John D. Merrifield of Victoria. Since 2003, the old launch has been owned by James G. Jack and Marilyn Jack of Vancouver, who renamed her *Sea Ox*. Today, the elegant yacht can often be seen around BC waters or tied up at local marinas.

Unknown rum runner held at the Vancouver Maritime Museum. Photo courtesy of the Vancouver Maritime Museum.

UNKNOWN

Another smaller launch, unfortunately unidentified, is in the collection of the Vancouver Maritime Museum after it was transferred to their collection from the British Columbia Provincial Museum in 1972. Twenty-one feet long with a long narrow hull and flared sharp bow with a "V" bottom, it was thought the fast hull was possibly a "Sang master" design of Vancouver. While one source has it that the launch had been involved in running liquor out of Vancouver and the Gulf Islands, Dr. Lorne Hammond, curator of history at the Royal BC Museum, suggested she may have been used as a rum runner working out of Sidney on Vancouver Island.

ACKNOWLEDGMENTS

There's always been this common misconception out among the reading public that being a writer is primarily a solitary endeavour. While this generally may hold true for fiction, it's certainly not the case for non-fiction and especially not for those of us intent on working up a comprehensive history book. Since I've been fascinated with our local maritime culture over the course of my life living here along the northwest coast of the Americas, and having dedicated myself to researching and writing about it for some thirty years now, this required countless hours spent in many a museum, archival collection and library.

Fortunately, having lived on Vancouver Island and, in particular, having grown up in Victoria, I've been particularly blessed with the fact that the British Columbia Archives is so readily accessible right next to our provincial museum. Over the years, I became one of the regulars and have gotten to know the institution's staff by first names. They always have been more than gracious running down particular records for me. Then there's our Maritime Museum of BC, which is temporarily and very conveniently located just the other side of the Empress Hotel from BC Archives. Here too the museum's library and archives have been manned by excellent staff. But just as important are their dedicated and knowledgeable volunteers, many retired with a background of working on the water most of their lives.

Meanwhile the Vancouver Maritime Museum (VMM) has also been blessed with particularly helpful staff and volunteers over the years. When I first walked in the door back in the late 1980s, I was met by the curator at the time, a most charming gentleman who proceeded to take me under his wing. Regardless that I had virtually no background in researching maritime history and was a stranger, Leonard McCann led me around the library and familiarized me with all its holdings. Len also was kind enough to reassure me that my project, identifying and researching the history of the fifteen scrapped ships beached at the Comox Logging & Railway Company's log dump at Royston, BC, to serve as a breakwater, was indeed a worthy project.

Just out the door from the VMM is the City of Vancouver Archives, which also proved a veritable treasure house. Sometime in the past, a collection of primary documents was delivered up to their collection by the Vancouver Maritime Museum (prior to the arrival of Leonard McCann). These include original logbooks and "Agreement or Articles and Account of Crew," the legal document that crew were required to sign before a ship departed on a voyage. From these, I was able to track the movements of the mother ship *Malahat* throughout her rum running years and, what was particularly exciting, check out the actual handwritten entries of crewmembers who signed on. Along with primary source material, many a photograph of a vessel that was involved in the rum running trade is held in our archives and museum collections and one can either inspect the original prints themselves at these facilities or, better yet, view them online. Also, another great source of early photos proved to be Vancouver Public Library's Special Collections. It should also be mentioned that the Greater Victoria Public Library, Alberni District Historical Society and both the Salt Spring Island and Oak Bay archive collections proved rewarding in running down particular threads of information or photographs.

Other institutions farther afield included the Prince Rupert City and Regional Archives, where I was able to access material on what was going down just a little to the northwest outside Alaskan waters. Meanwhile, just across the line in Seattle is the Puget Sound Maritime Historical Society, of which I've been a member for some twenty years, home to a very extensive collection of coastal maritime history. But where I really struck pay dirt was in San Francisco. What with it being a major Pacific port going back to the 1849 gold rush years, it is home to what is probably one of the best maritime history centres on the planet, the J. Porter Shaw Library down on the city's maritime park. But beside that, there's the city's public library just off Market Street, which is home to San Francisco's History Center. Here I had the most enjoyable time scanning microfilm of the *San Francisco Chronicle* and *Examiner* newspapers, whose reporters kept the American public very much entertained with news of the latest capture of a Canadian mother ship off the California coast and the subsequent trial of its owners, captains and crew.

Overall, newspapers of the day proved a particularly valuable resource and I spent many an hour flipping through microfilm of the *Vancouver Sun* and *Province* and Victoria's *Daily Times* and *Colonist* in the Victoria Public Library and down at BC Archives. By the early 1920s, a particularly excellent source of information proved to be the *Colonist*'s Marine & Transportation page. With its updates on the arrival of large freighters from Scotland, England and Antwerp loaded down with the finest in Scotch, rye and champagne as well as the latest news on the mother ships sitting off the Oregon and California coast, the message was made most apparent that British Columbia's liquor export industry was generally an above-board and legal undertaking during the US Prohibition years.

A special mention should be made of an incredible online resource easily accessible to all who have an interest in learning about British Columbia's floating heritage. This is the fabulous website created by John MacFarlane, nauticapedia.ca. It was here I would often begin my investigations and was able to get a good handle on where and when a particular vessel was built, its specifications and the various owners whose hands she passed through over the years. With this rudimentary information, I could work on chasing down both primary and secondary documentation and fill in the story of a vessel's years in the trade more efficiently.

But above all, I can't stress enough how deeply indebted I am to that handful of souls who finally put aside the brotherhood's creed, "Don't never tell nobody nothing, nohow" later in life and opened up to share their stories. Fortunately, along with Fraser "Sparky" Miles's autobiography, Hugh Garling wrote a series of articles on rum running which ran in *Harbour & Shipping* magazine from late 1988 through early 1991; Jim Stone worked up a biography of his father, Captain Stuart Stone; while Marion Parker and Robert Tyrrell did a fine job of getting rum runner Johnny Schnarr's most entertaining stories of his adventures out in book form.

Sparky Miles's son, Jim, deserves recognition too, for being particularly generous in sharing his father's photo collection with this writer. Then there was another of my mentors, Fred Rogers, who spent many years of his life diving and recording countless shipwrecks along the entire coast of BC, and then went to some effort to work up his findings and publish them all in two books: *Shipwrecks of British Columbia* and *More Shipwrecks of British Columbia*. I'm especially indebted to Fred since it was he who gifted me his entire collection of old *Harbour & Shipping* magazines which just so happened to include Hugh Garling's articles. Another gentleman who was also of particular help with Hugh Garling's photos was Bill Singer, owner of the Rumrunner Pub, located down along the beachfront of Sidney, BC, which just happens to overlook Haro Strait, an active marketplace during the rum running years. Then there was my good bud Doug Kerr, who devoted many hours to cleaning up many of the old photos presented in this book. Another individual who was generous sharing photographs was Valerie Allen, who forwarded the photo collection belonging to her grandfather Fred Sailes, an engineer aboard *Malahat*.

Surprisingly, the first one to open up and share his take on all that went down was none other Captain Charles Hudson, managing director of Consolidated Exporters Corporation Limited, the big liquor export consortium that operated out of Vancouver while

Prohibition remained in force throughout the US. Captain Hudson was most gracious and very open to being interviewed by Imbert Orchard and Ron Burton, and today these oral history tapes are available at the BC Archives. And as for the actual participants themselves, who were crewing aboard the mother ships, fast shore boats, packers and basically anything that floated that they could make a good dollar with, a handful stepped forward in later years and allowed themselves to be interviewed and recounted their most fascinating adventures.

My apologies to all those others who provided me with snippets of information or other leads worth checking out, but to credit them all would have required tacking on another page or two to my acknowledgments. Still, special mention should be made of Ken Gibson in Tofino who kept me well informed about a number of the characters over on the west coast of the island who became quite prominent in the trade; my mentor, Frank Clapp, who is always a stickler for primary documentation; and Ron Greene, a most diligent researcher and writer who has served as my man on the ground in the BC Archives. And, last but certainly not the least, my partner, Paula Wild, who has faithfully served as my in-house advisor, mentor and editor over the years.

"History ... is not a re-creation of the past. It's an assessment of the past based on documents provided by people in archives and museums who will answer your letters."
— Stephen C. Levi

SOURCES

BOOKS

Barman, Jean. *The West Beyond the West: A History of British Columbia.*
Toronto: University of Toronto Press, 1991

Campbell, Robert A. *Demon Rum or Easy Money: Government Control of Liquor in British Columbia from Prohibition to Privatization.* Ottawa: Carleton University Press, 1991

Canney, Donald L. *U.S. Coast Guard and Revenue Cutters, 1790–1935.* Annapolis, MD: Naval Institute Press, 1995

Clark, Cecil. *The Man Who Was Hanged by a Thread and Other Tales from BC's First Lawmen.* Victoria: Heritage House, 2011

Faith, Nicholas. *The Bronfmans: The Rise and Fall of the House of Seagram.* New York: St. Martin's Press, 2007

Francis, Daniel, ed. *Encyclopedia of British Columbia.* Madeira Park, BC: Harbour Publishing, 2000

Gibbs, Jim. *Shipwrecks off Juan de Fuca.* Portland, OR: Binford and Mort, 1968

Gray, James H. *Booze: The Impact of Whiskey on the Prairie West.* Toronto: Macmillan Canada, 1972

Greene, Ruth. *Personality Ships of British Columbia.* West Vancouver: Marine Tapestry Publications, 1969

Hacking, Norman. *The Two Barneys: A Nostalgic Memoir About Two Great British Columbia Seamen.* Vancouver: Gordon Soules Book Publishers, 1984

Henderson, Maxwell. *Plain Talk! Memoirs of an Auditor General.* Toronto: McClelland & Stewart, 1984

Horsfield, Margaret, and Ian Kennedy. *Tofino and Clayoquot Sound: A History.* Madeira Park, BC: Harbour Publishing, 2014

Irving, Joseph. *Supplement to the Annals of our Time: A Diurnal of Events Social and Political, Home and Foreign From July 22, 1878, to the Jubilee, June 20, 1887*. London and New York: MacMillan and Co., 1889

MacFarlane, John M. *The Nauticapedia List of British Columbia's Floating Heritage: Volume 1*. Delta, BC: John M. MacFarlane, 2014

MacFarlane, John M. *The Nauticapedia List of British Columbia's Floating Heritage: Volume 2*. Delta, BC: John M. MacFarlane, 2014

Macpherson, Ken, and John Burgess. *The Ships of Canada's Naval Forces 1910– 1993: A Complete Pictorial History of Canadian Warships*. St. Catherines, ON.: Vanwell Publishing, 1994

McIntosh, Dave. "Chapter V: The Customs Scandal of 1926," *The Collectors: A History of Canadian Customs and Excise*. Toronto: NC Press in association with Revenue Canada, Customs and Excise and the Canadian Government Publishing Centre, Supply and Services Canada, 1984

Metcalfe, Philip. *Whispering Wires: The Tragic Tale of an American Bootlegger*. Portland, OR: Inkwater Press, 2007

Miles, Fraser. *Slow Boat on Rum Row*. Madeira Park, BC: Harbour Publishing, 1992

Morris, Rob. *Coasters: Uchuck III, Lady Rose, Frances Barkley, Tyee Princess*. Victoria: Horsdal & Schubart Publishers, 1993

Murray, Peter. *The Vagabond Fleet: A Chronicle of the North Pacific Sealing Schooner Trade*. Victoria: Sono Nis Press, 1988

Newell, Gordon, ed. *The H. W. McCurdy Marine History of the Pacific Northwest*. Seattle: Superior Publishing, 1966

Newman, Peter C. *Bronfman Dynasty: The Rothschilds of the New World*. Toronto: McClelland and Stewart, 1978

Nicholson, George. *Vancouver Island's West Coast: 1762–1962*. Victoria, BC: Morriss Printing Company, 1965

O'Hara, Tracy, and Bent Sivertz. *The Life of Bent Gestur Sivertz: A Seaman, a Teacher, and a Worker in the Canadian Arctic*. Victoria: Trafford Publishing, 2000

Parker, Marion, and Robert Tyrrell. *Rumrunner: The Life and Times of Johnny Schnarr*. Victoria: Orca Book Publishers, 1988

Schneider, Stephen. *Iced: The Story of Organized Crime in Canada*. Mississauga, ON: John Wiley & Sons Canada, 2009

Skalley, Michael. *Foss: Ninety Years of Towboating*. Seattle: Superior Publishing, 1981

Stone, Jim. *My Dad, The Rum Runner*. Waterloo, ON: North Waterloo Academic Press, 2002

Stonier-Newman, Lynne. *Policing a Pioneer Province: The BC Provincial Police 1858–1950*. Madeira Park, BC: Harbour Publishing, 1991

Taylor, G. W. *Shipyards of British Columbia: The Principal Companies*, Vancouver: J. J. Douglas, 1986

Verrill, A. Hyatt. *Smugglers and Smuggling*. New York: Duffield and Company, 1924

White, Gary M. *Hall Brothers Shipbuilders*. San Francisco: Arcadia Publishing, 2008

Willoughby, Malcolm F. Commander USCGR(T). *Rum War at Sea*. Washington, DC: Treasury Department, United States Coast Guard, 1964

Wishart, Bruce. *Charlie's Tugboat Tales: Charlie Currie's Waterfront Tales from the British Columbia Coast*. Prince Rupert, BC: Prince Rupert This Week, 1997

Wright, E. W., ed. *Lewis & Dryden's Marine History of the Pacific Northwest*. New York: Antiquarian Press, 1961

ARCHIVAL RECORDS

"Articles of Association of 'Arctic Fur Traders Exchange Limited.'" Companies Act, 1921, May 31, 1922, Number 6404, microfilm B05160, GR-1526: British Columbia Registrar of Companies: Files of Dissolved Companies, British Columbia Archives

Canada. "Report of the Special Committee investigating the Administration of the Department of Customs and Excise." 2 vols. Ottawa, 1926. Canada. "Royal Commission on Customs and Excise, Interim Reports 1 to 10 and Final Report." Ottawa, 1928

"Canadian-Mexican Shipping Company." File 5608, microfilm B05202, GR-1526: British Columbia Registrar of Companies: Files of Dissolved Companies, British Columbia Archives

Catlow, Henry. Letter dated October 9, 1926, from Henry Catlow, Special Undercover Customs agent, to A. J. Gaudron, Superintendent, Director of Criminal Investigations, RCMP, for Customs Department (letter No. 916223) in file: "Operations of Mr. A. J. Gaudron, Superintendent of Criminal Investigation RCMP re Border Patrol and Prevention of Smuggling." National Archives of Canada, RG16: Records of the Department of National Revenue, Vol. 789, File No. 128256, Letter No. 916223

City of Vancouver Archives. AM54- Major Matthews Collection, Box 505-B-4, File 50, Memo of conversation with Mr. George C. Reifel, Vancouver, BC, January 31, 1938.

Correspondence File of the Vessel *Morris*, National Archives, Washington, DC, 1932

Dominions Office Letters Received, no. 640 ---? for the Secretary of State to Sir W. L. Allardyce, August 11, 1924, with treaty enclosure; and (illegible) Dominions Office to Sir W. L. Allardyce, November 17, 1924. GN 1/2/0, Provincial Archives of Newfoundland and Labrador

"List of Vessels of the Revenue Cutter Service," *Thirty-Seventh Annual List of the Merchant Vessels of the United States*. Washington, DC: Department of Commerce and Labor, Government Printing Office, 1905, pp. 441–42

Lloyd's of London. *Lloyd's Register of British and Foreign Shipping*. London: Lloyd's of London

"Memorandum of Association of Consolidated Exporters Corporation Limited." Companies Act, 1921, August 25, 1922, Number 6545, microfilm B05204, GR-1526: British Columbia Registrar of Companies: Files of Dissolved Companies, British Columbia Archives

"South Seas Traders Limited Certificate of Incorporation." Companies Act, Number 12,607, microfilm B05313, GR-1526: British Columbia Registrar of Companies: Files of Dissolved Companies, British Columbia Archives

United States Coast Guard Treasury Department, Oakland, California. Correspondence 4 January 1933 to Commander, Section Base 11

NEWSPAPER AND MAGAZINE ARTICLES

Clark, Cecil. "Hideous Crimes Remain Unsolved and Unpunished: Murder Sometimes Baffles Science," *The Daily Colonist Islander*, Victoria: August 18, 1963, pp. 6–7

Clark, Cecil. "Victoria's Rum Ship Made $1,000,000 Trips," *The Daily Colonist Islander*, Victoria: February 9, 1969, pp. 12, 16

Cooke, Alfred Edward. Bureau of Statistics figures cited in "The Canadian Liquor System I – Evils of Government Control," *The New York Times Current History*, October 1929, pp. 64–73

Crawford, Richard. "Imported Illegality Poured in during Prohibition," *San Diego Union-Tribune*, San Diego: August 7, 2008

Cuthbert, Lori. "State of the Nation," *Vancouver Sun*, Vancouver: October 15, 2016, NP1-3

"Estimates of the Smuggling Situation to the Repeal of the 18th Amendment," United States Coast Guard Intelligence Report. United States National Archives.

Frisbie, Robert Dean. "Rum Row: Western," *American Mercury*, May 1932, vol. 26, pp. 62–68

Garling, Hugh, "Rum Running on the West Coast: A Look at the Vessels and People," columns in *Harbour & Shipping*, Vancouver, November 1988–July 1991

Gibson, Ken. "Emmie May Binns—1900–1994," unpublished manuscript

Greene Bailey, Ruth. "Rum Runners Captured: Marine Notebook," *Harbour & Shipping*, Vancouver, May 1964, pp. 168, 260–62

"Harbour & Shipping News Dragnet"; "Mosquito Fleet Notes"; "Vessels Registered at Port of Vancouver," *Harbour & Shipping*, Vancouver, 1929–33

Hesse, Jurgen. "Cappy Hudson Rumrunner!" Vancouver, *Western Living*, May 1976, pp. 6–9

Hyman, Frank. "The Story of Rum Running," *Historic Writings*. Los Angeles: Panpipes Press, 1966, pp. 22–31

Kaplan, H. R. "A Toast to the "Rum Fleet," *Naval Institute Proceedings* 94, May 1968, pp. 85–90

Kelly, L. V. "The Ships that Clear 'Deep-Sea.'" *Canadian Merchant Service Guild Yearbook 1931*, pp. 38–42

Kelly, L. V. "The Stodgy Malahat Gave ...," *The Vancouver Province B.C. Magazine*, March 12, 1955, pp. 4–5

Kelly, L. V. "Vancouver's Rum-Running Flotilla ..." *The Sunday Province* (Vancouver)

Kreiger, Dan. "If These Logs Could Talk: Prohibition in SLO County: Times Past," *The Tribune*, San Luis Obispo, CA: February 17, 2002, p. B5

"Liquor and Liberals: Patronage and Government Control in British Columbia, 1920–1928," *BC Studies*, No. 77, Spring 1988, pp. 30–53

"Liquor Cargo Held at Coast is Dwindling," *The Citizen* (Ottawa), January 6, 1927, p. 4

"Liquor Treaty: Britain and USA," *The Brisbane Courier*, Brisbane, Australia: May 24, 1924, p. 17

Lonsdale, Captain A. L., US Coast Guard. "Rumrunners on Puget Sound," *American West*, vol. 9, issue 6, November 1972

Luce, P. W. "The Odd Angle," *The Vancouver Daily Province*, circa March 5, 1927

Lund, John. "Queen of Rum Row," *Islander*, supplement to *Times Colonist* (Victoria), September 11, 1994, p. 1

Lund, John. "Two Queens of Rum Row: Rum Runners of the Northwest," *Pacific Yachting*, September 2007, pp. 50–55

Lyndell, Honoree B. "Chasing the Rum Runners—Adventures of a British Columbia Coast Guard Boat," *The Sunday Province* (Vancouver), March 4, 1928, p. 11

MacKenzie, D. O. "SS PRINCE ALBERT – Rum Runner." *The Bulletin: Quarterly Journal of the Maritime Museum of British Columbia*, Spring 1980, pp. 4–8, 16

McInnis, Edgar. "The Political Aspect of Whisky." *Canadian Forum*, September 1929, p. 414

Munday, Don, "Hijacking of the "Beryl G," *The Shoulder Strap*, Winter Edition, 1940, pp. 65–76

"New Revenue Cutter Arcata Will Engage in Harbour Duty," *San Francisco Call*, San Francisco: January 7, 1903, p. 10

Nicholson, George. "Gallant Old Quadra, Chore-Boy of the Coast," *The Daily Colonist Islander* (Victoria), May 24, 1964, pp. 10–11

O'Leary, M. Grattan. "Thar's Gold in Them Thar Stills," *Maclean's*, September 1, 1932, pp. 15, 52–53

Olsen, M. P. Letter to Charles Defieux, Marine Editor, *Vancouver Sun*, February 15, 1966.

"Rum-Runners are shot down," unidentified newspaper clip, June 28, 1924

The Seamen's Journal: San Francisco: International Seamen's Union of America; Sailors' Union of the Pacific, November 1927, p. 342

Sannah, Kelefa. "Drunk with Power: What Was Prohibition Really About?" *The New Yorker*, December 21 & 28, 2015, pp. 105–10

"Schooner 'Malahat' Sold," Steamship & Port News. *Harbour and Shipping*, May 1923, p. 26

Sivertz, Bent Gestur. "Rumrunning Repairs," *The Bulletin: Quarterly Journal of the Maritime Museum of British Columbia*, Summer 1980, pp. 7–9

Starkins, Ed. "Rum Running," *Raincoast Chronicles First Five: Collector's Edition*. Madeira Park, BC: Harbour Publishing, 1977, pp. 15–17

Stott, Art. "Yo-ho-ho and a Runner of Rum: On the Times." *Victoria Daily Times*, July 15, 1964

Sutherland, Mrs. S., ed. "Smuggling in Shetland in the 19th Century: from the notes of William Smith, Sandwick, 1872–1949." *Shetland Life*, Parts One to Three, March, April, May 1992

"The Profitable Pursuit of Rum-Running over the Canadian Border," *Literary Digest*, October 16, 1920, p. 65

"United Distillers, Ltd.: New Company & Company News," *Harbour & Shipping*. Vancouver: 1925

Waters, Harold, "How We Scuttled Rum Row," *Popular Boating*, April 1962, pp. 116–18, 170, 172

Wrotnowski, Dorothy. "Island's Emmie May Sailed on Great Ship," *Daily Colonist* (Victoria), March 7, 1971, p. 24

NEWSPAPERS

British Columbia Archives newspaper microfilm file, D-19 123, #2784-85

The Daily Colonist, Victoria: 1916–1933

The Daily Colonist Islander, Victoria: 1963

The Daily Colonist, Marine and Transportation page, Victoria: 1922–33

San Francisco Chronicle: 1925–1928

San Francisco Examiner: 1928

The Vancouver Daily Province: 1927–1933

The Vancouver Sun: 1923–1933

Victoria Daily Times: 1922–1927

INTERVIEWS, CONVERSATIONS AND PERSONAL CORRESPONDENCE

Valerie Allen, Hugh "Red" Garling, Ken Gibson, Ron Greene, Jim Kirkland, David Lansdowne, John MacFarlane, Jacques Marc, Jim Miles, Bill Singer, Bent Sivertz, M. Thompson

ORAL HISTORY TAPES

Branca, Angelo. Itter & Marlatt interview recorded 1977. British Columbia Archives: T2619:0003, p. 14 of transcript

Bittancourt, Len. CBC interview recorded October 8, 1965. Imbert Orchard records. British Columbia Archives: T0798:0001

Gisborne, Robert. Susan Winifred Mayse oral history collection, recorded March 21, 1984. British Columbia Archives: T4136:0001, T4136:0002

Hudson, Charles. Imbert Orchard 1960s interview. British Columbia Archives: T4255:0039

Hudson, Charles. Ron Burton "Rum-running interviews: tape 1." British Columbia Archives: T4407:0001

MARITIME HISTORY RESOURCES

Department of Commerce and Labour. *Wergeland*: "Fiftieth Annual List of the Merchant Vessels of the United States," Washington, DC: Government Printing Office, 1918

Department of Marine, Canada. *Speedway*: "Return for Wreck Register, 1925," RG 42, C-3-B, vol. 676. Library and Archives Canada

Department of Marine and Fisheries, Canada. "Agreement or Articles and Account of Crew," File 7, Boxes 88-0056 & 91-4210, GR 0215.1806, British Columbia Archives

Department of Marine and Fisheries, Canada. *List of Vessels on the Registry Books of the Dominion of Canada* (also known as the "Blue Book"), Ottawa: King's Printer, 1923–36

Department of Transport, Canada. *List of Vessels on the Registry Books of the Dominion of Canada*, Ottawa: Edmond Cloutier, Printer to the King's Most Excellent Majesty, 1941

Lloyd's of London. *Lloyd's Register of British and Foreign Shipping: Steamers & Motorships*. London: 1890–1936

Mercantile Marine Shipping Office, Department of Marine and Fisheries, Canada. *Malahat*: "Log Books" and "Agreement or Articles and Account of Crew," AMS-S1–, Vancouver Maritime Museum collection, City of Vancouver Archives

Mercantile Marine Shipping Office, Department of Marine and Fisheries, Canada. *Malahat*: "Log Books" and "Agreement or Articles and Account of Crew," File 1, Box 44; Files 22, 24, 27, 28, 33, 44, Boxes 411–67, GR 0215, British Columbia Archives

Registrar General of Shipping and Seamen, Great Britain. "Agreement and Account of Crew and Official Logbooks for British Empire Vessels, 1857–1942." Maritime History Archive, Memorial University of Newfoundland

Ryou I: Blue Book (official logbook of owner). David Lansdowne collection

Transport Canada. *Vancouver Ship Registry, Vessels Registered in Vancouver, 1913–33*

Transport Canada. *Victoria Ship Registry, Vessels Registered in Victoria, 1902–33*

ONLINE RESOURCES

Department of Justice. "Report Concerning a Conspiracy to Smuggle Whiskey into Oregon from Ships Off the Coast," *National Archives*, File Unit: Investigation of Sterling Traders, 1927–1933, Record Group 56: General Records of the Department of the Treasury, 1789–1990. National Archives Pacific Alaska Region Catalog, research.archives.gov/id/298434

Francis, Daniel. "Prohibition in BC," *KnowBC*, November 12, 2012, knowbc.com /Knowbc-Blog/Prohibition-in-BC

Harvey, Robert. "Stadacona – Yacht, Rum Runner and Naval Vessel," *Nauticapedia*, 2013, www.nauticapedia.ca/Articles/Stadacona.php

Justia. "United States v. Ferris et al. (two cases). Same vs. Stone," *Justia: US Law*, law.justia.com/cases/federal/district-courts/F2/19/925/1502205/

Library of Congress. "Temperance and Prohibition," *The Library of Congress*, www.memory.loc.gov/ammem/today/oct28.html

Lopes, Emmanuel. "The Bittancourts of Salt Spring Island." *Portuguese Pioneers of BC*, August 24, 2011, portuguesepioneersofbc.blogspot.ca /2011/08/bittancourts-of-salt-spring-island-from.html

MacFarlane, John M. "The Nauticapedia Project Vessels Database," *Nauticapedia*, www.nauticapedia.ca/dbase/Query/dbsubmit_Vessel.php

McClary, Daryl C. "Olmstead, Roy (1886–1966) – King of King County Bootleggers," *The Free Encyclopedia of Washington State History*, November 13, 2002, www.historylink.org/File/4015

NavSource Naval History. "USCGC Arcata (CG 11)," *NavSource Naval History: Photographic History of the U.S. Navy*, www.navsource.org/archives/12 /179913.htm

Pennington, Christine. "Mixing Business and Pleasure: Rum-Running in Vancouver, British Columbia," *Histories of British Columbia*, blogs.ubc.ca/ hist305/histories-of-bc/rum-running-in-vancouver/

Roberts, Stephen S. "Class: CAESAR (AC-16)," *ShipScribe*, www.shipscribe.com /usnaux/AC/AC16.html

Southampton City Council. "Federalship," *Plimsoll ShipData*, www.plimsoll shipdata.org/ship.php?ship_id=40738&name=Federalship

Supreme Court of the United States. "Ford et al. v. UNITED STATES," *Cornell Law School*, www.law.cornell.edu/supremecourt/text/273/593

"The Henry Reifel & Annie Elizabeth Brown Family," *Reifel Family Tree*, reifeltree.tripod.com/henryanniereifel.html

The Historic Chief Skugaid, www.chiefskugaid.org

"United States Coast Guard Historian's Office," *United States Coast Guard*, www.history.uscg.mil/

Vancouver Public Library. *British Columbia City Directories, 1860–1955*, bccd.vpl.ca/

Vanderhill, Jason. "Illustrated Vancouver Vol. 6 – United Distillers Ltd., Vancouver," *Vancouver is Awesome*, www.vancouverisawesome.com /2012/07/22/illustrated-vancouver-vol-26-united-distillers/

Woofenden, Todd. *The Subchaser Archives*, www.subchaser.org

INDEX

Page numbers in **bold** indicate illustrations